BIOTECHNOLOGY IN AGRICULTURE SERIES

General Editor: Gabrielle J. Persley, Biotechnology Adviser, Environmentally Sustainable Development, The World Bank, Washington DC, USA.

For a number of years, biotechnology has held out the prospect for major advances in agricultural production, but only recently have the results of this new revolution started to reach application in the field. The potential for further rapid developments is, however, immense.

The aim of this book series is to review advances and current knowledge in key areas of biotechnology as applied to crop and animal production, forestry and food science. Some titles focus on individual crop species, others on specific goals such as plant protection or animal health, with yet others addressing particular methodologies such as tissue culture, transformation or immunoassay. In some cases, relevant molecular and cell biology and genetics are also covered. Issues of relevance to both industrialized and developing countries are addressed and social, economic and legal implications are also considered. Most titles are written for research workers in the biological sciences and agriculture, but some are also useful as textbooks for senior-level students in these disciplines.

BIOTECHNOLOGY IN AGRICULTURE SERIES

Titles Available:

1: Beyond Mendel's Garden: Biotechnology in the Service of World Agriculture *
G.J. Persley

2: Agricultural Biotechnology: Opportunities for International Development
Edited by G.J. Persley

3: The Molecular and Cellular Biology of the Potato *
Edited by M.E. Vayda and W.D. Park

4: Advanced Methods in Plant Breeding and Biotechnology
Edited by D.R. Murray

5: Barley: Genetics, Biochemistry, Molecular Biology and Biotechnology
Edited by P.R. Shewry

6: Rice Biotechnology
Edited by G.S. Khush and G.H. Toenniessen

7: Plant Genetic Manipulation for Crop Protection *
Edited by A. Gatehouse, V. Hilder and D. Boulter

8: Biotechnology of Perennial Fruit Crops
Edited by F.A. Hammerschlag and R.E. Litz

9: Bioconversion of Forest and Agricultural Plant Residues
Edited by J.N. Saddler

10: Peas: Genetics, Molecular Biology and Biotechnology
Edited by R. Casey and D.R. Davies

11: Laboratory Production of Cattle Embryos
I. Gordon

12: The Molecular and Cellular Biology of the Potato, 2nd edn
Edited by W.R. Belknap, M.E. Vayda and W.D. Park

13: New Diagnostics in Crop Sciences
Edited by J.H. Skerritt and R. Appels

14: Soybean: Genetics, Molecular Biology and Biotechnology
Edited by D.P.S. Verma and R.C. Shoemaker

15: Biotechnology and Integrated Pest Management
Edited by G.J. Persley

16: Biotechnology of Ornamental Plants
Edited by R.L. Geneve, J.E. Preece and S.A. Merkle

17: Biotechnology and the Improvement of Forage Legumes
Edited by B.D. McKersie and D.C.W. Brown

18: Milk Composition, Production and Biotechnology
Edited by R.A.S. Welch, D.J.W. Burns, S.R. Davis, A.I. Popay and C.G. Prosser

19: Biotechnology and Plant Genetic Resources: Conservation and Use
Edited by J.A. Callow, B.V. Ford-Lloyd and H.J. Newbury

20: Intellectual Property Rights in Agricultural Biotechnology
Edited by F.H. Erbisch and K.M. Maredia

21: Agricultural Biotechnology in International Development
Edited by C. Ives and B. Bedford

22: The Exploitation of Plant Genetic Information: Political Strategies in Crop Development
R. Pistorius and J. van Wijk

23: Managing Agricultural Biotechnology: Addressing Research Programs Needs and Policy Implications
Edited by J.I. Cohen

* Out of print

The Exploitation of Plant Genetic Information

Political Strategies in Crop Development

Robin Pistorius

Department of Political Science
Universiteit van Amsterdam
Oudezijds Achterburgwal 237
1012 DL Amsterdam
The Netherlands

and

Jeroen van Wijk

Department of Public Management
Faculty of Business Administration
Erasmus Universiteit Rotterdam
PO Box 1738
3000 DR Rotterdam
The Netherlands

CABI *Publishing*

CABI *Publishing* is a division of CAB *International*

CABI Publishing
CAB International
Wallingford
Oxon OX10 8DE
UK

Tel: +44 (0)1491 832111
Fax: +44 (0)1491 833508
Email: cabi@cabi.org

CABI Publishing
10 E 40th Street
Suite 3203
New York, NY 10016
USA

Tel: +1 212 481 7018
Fax: +1 212 686 7993
Email: cabi-nao@cabi.org

A catalogue record for this book is available from the British Library, London, UK.

Library of Congress Cataloging-in-Publication Data
Pistorius, Robin.
The exploitation of plant genetic information : political strategies in crop development / Robin Pistorius, Jeroen van Wijk.
p. cm.
Includes bibliographical references.
ISBN 0-85199-363-X (alk. paper)
1. Germplasm resources, Plant. 2. Crops--Germplasm resources.
I. Wijk, Jeroen van. II. Title.
SB123.3.P565 1999
333.95'34--dc21 99-16220
CIP

ISBN 0 85199 363 X

Printed and bound in the UK at the University Press, Cambridge, from copy supplied by the author.

To

Luuk, Tim, and Winnie

Contents

Abbreviations

AIPPI	International Association for the Protection of Intellectual Property
ASSINSEL	International Association of Plant Breeders for the Protection of Plant Varieties
ASTA	American Seed Trade Association
CBD	Convention on Biological Diversity
CEO	Chief executive officer
CGIAR	Consultative Group on International Agricultural Research
CIAT	International Center for Tropical Agriculture (CGIAR)
CIMMYT	International Center for the Improvement of Maize and Wheat (CGIAR)
CORPOICA	Corporación Colombiana de Investigación Agropecuario (Colombia)
CPGR	Commission on Plant Genetic Resources (FAO)
DNA	Deoxyribonucleic acid
EEC	European Economic Community
EPC	European Patent Convention
EPO	European Patent Office
FAO	Food and Agriculture Organization of the United Nations
GATT	General Agreement on Tariffs and Trade
GEF	Global Environment Facility
IARC	International agricultural research centre
IBPGR	International Board for Plant Genetic Resources (now IPGRI)
ICA	Instituto Colombiano Agropecuario (Colombia)
ICDA	International Coalition for Development Action
INIA	Instituto Nacional de Investigación Agropecuaria (Chile)
IPGRI	International Plant Genetic Resources Institute (CGIAR)
IPR	Intellectual property rights

IRRI	International Rice Research Institute (CGIAR)
IUCN	World Conservation Union
JICA	Japanese Agency for International Cooperation
KWI	Kaiser-Wilhelm-Institut (Germany)
LAMP	Latin American Maize Program
MAP	Mexican Agricultural Program
MERCOSUR	Mercado Común del Sur
NAFTA	North American Free Trade Agreement
NARS	National agricultural research system
NBF	New biotechnology firm
NIEO	New International Economic Order
NGO	Non-governmental organization
NPGS	National Plant Germplasm System (USA)
NSSL	National Seed Storage Laboratory (USA)
NTC	Non-traditional commodity
OECD	Organisation for Economic Co-operation and Development
OPV	Open-pollinated variety
PBR	Plant breeders' rights
PGR	Plant genetic resources
PPA	Plant Patent Act (USA)
PVP	Plant variety protection
TBA	Technical Board of Appeal (EPO)
TNE	Transnational enterprise
TRIPS	Trade-Related Aspects of Intellectual Property Rights
UNCED	United Nations Conference on Environment and Development
UNEP	United Nations Environmental Programme
UPOV	International Union for the Protection of New Varieties of Plants
USDA	U.S. Department of Agriculture
USPTO	U.S. Patent and Trademark Office
VIR	All-Union Institute of Plant Industry (USSR)
WBCSD	World Business Council for Sustainable Development
WTO	World Trade Organization

Figures

Preface

The time that plant genetic information was dealt with exclusively in scientific circles is definitely over. Business people, politicians, lawyers, environmentalists, farmers, indigenous communities, and consumers have also involved themselves in the exploitation and conservation of genetic information. To them, genetic information does not only refer to DNA, but is also related to food security, health care, cultural survival, new markets, or intellectual property rights. The stakes are high, and this explains why the exploitation of genetic information has become a controversial issue in international politics. Scientists in the crop development industry can no longer operate in a quiet corner of society. The many new stakeholders have formulated views on how genetic information should be exploited and are lobbying forcefully for the implementation of their views.

For quite a number of years we have been fascinated by the attempts to regulate the exploitation of genetic information and the international conflicts that were triggered by these attempts. We have reported on new developments in the *Biotechnology and Development Monitor*, a journal produced at the Department of Political Science of the University of Amsterdam, which has been our base for the past ten years. We have also been involved in various academic and policy-oriented projects related to the conservation and intellectual property protection of genetic information. Over the years, this work inspired us to start a more integral research project in which conservation and plant-related intellectual property protection issues are framed within their most obvious context, i.e. that of agricultural politics.

A great deal of the motivation to write this book came from what we increasingly considered as shortcomings in the existing literature on conservation and intellectual property protection. First, there is the problem of disciplinary orthodoxy. Academic discussions on regulating the conservation and protection of genetic information seem to take place within the various disciplines involved, while there is relatively little exchange of ideas and knowledge among them. Plant breeders tend to perceive the issue basically from a biological and agronomic point of view; lawyers follow the legal path; molecular biologists have a biotechnological approach; economists focus on the cost and benefits of mining genetic information as an industrial resource; and political scientists are preoccupied with national and international power relations. It was our intention to navigate between these disciplines and to unlock and combine the valuable knowledge from all the various sources. Nevertheless, we have always been aware of our 'guest' status in other disciplines. We had the ambition to write a *political* analysis of the exploitation of plant genetic information, which implied an analysis of the power constellations behind apparently neutral technological, economic and legal developments.

Another motivation to write this book was an ambition to offer the reader some tools to take the plunge into the highly complex conflict on genetic information. We find most social science contributions on the issue unsatisfactory, as they explain the controversies surrounding genetic information in terms of a simple bi-polar confrontation between the "North" and the "South". We attempt to present the conflict in terms of clashes between different types of agriculture. Actors involved are not just Northern or Southern governments, but a whole range of interest groups, such as governmental departments, business associations, farm and consumers' organizations, environment-oriented NGOs, and indigenous people's organizations. As the reader will notice, alliances and conflicts between these groups are far more complex and fluid than the intergovernmental or hemispheric blocks on which most analyses rely.

Many people inspired, assisted, or in some other way enabled us to complete this work. Very much appreciated have been the comments on earlier versions of parts of this book from Jos Bijman, Jean Carrière, Don Duvick, Huib Ghijsen, Henk Hobbelink, Gerd Junne, Niels Louwaars, John Komen, Volker Lehmann, C. Mastenbroek, Dwijen Rangekar, Hans Raven, Rob van Tulder, Bert Visser, and two anonymous reviewers. A part of the interviews was carried out in collaboration with or with the assistance of Félix Moreno, Rodolfo Alvarado, and Marta Rueda in Colombia, and with Verónica Paeile in Chile. Michael Blass did some very valuable text editing.

Finance for this research was provided by the Centre for Study and Documentation of Latin America (CEDLA), the University of Amsterdam, the Netherlands Organization for Scientific Research (NWO), and the J.E. Jurriaanse Stichting.

Amsterdam, September 1999

Introduction

Biological diversity is often thought to be an area exclusively for biologists and environmentalists. In the past 20 years they have indeed played a leading role in the discussion on the conservation of 'biodiversity'. Since the United Nations Conference on Environment and Development (UNCED) in 1992, the loss of biodiversity has obtained a prominent place in the debate on the global environment. The overall message is that there are 30 million species left, but every day another one is lost forever.

Less well known is that the environmental/biological interest in maintaining biodiversity is increasingly shared by industry, albeit from a more utilitarian perspective. New biotechnological techniques have enabled industrial corporations to screen the earth's plants, insects, and microorganisms at unprecedented speed for useful genetic information. Genes that are economically interesting can now be 'engineered' and transferred from one organism to another, for medical or agricultural purposes, or a combination of the two. Biodiversity has thus become a potential industrial, genetic resource.

Because of the industrial interest in genetic information, the realization is widening in that ordinary plants, insects, microorganisms, or human blood cells represent economic value. This awareness formed the starting point for a long, intense and still ongoing conflict over who is to benefit from the exploitation of genetic information. The conflict has become most visible in political debates between industrialized and developing countries, most often at the United Nations forum. Plant genetic resources are found especially in relatively undisturbed areas in developing countries, while the technological know-how to exploit organisms as a resource is primarily available in advanced OECD countries._

The first clashes took place within the FAO in the early 1980s, revived in the 1990s during the preparations for the UNCED in 1992, and resulted in the U.S. Congress refusing to sign the Convention on Biological Diversity (CBD). Political turmoil of the same magnitude occurred during the Uruguay Round of multilateral trade negotiations, when an international agreement on Trade-Related Aspects of Intellectual Property Rights (TRIPS) was being prepared. In the light of the heated debate surrounding the implementation of the CBD and the TRIPS agreement, it can be expected that controversies on the exploitation of genetic information will last well into the 21st century.

Although most conservation and intellectual property rights (IPR) regulation is developed in OECD countries, the implementation also takes place in developing countries. 'Global' conservation regulation is required to protect forests, coastal areas, wetlands and farmlands both for long-term ecological and for industrial interests. 'Global' IPR regulation is designed to prevent farmers and breeders from using plant varieties without paying royalties. IPR protection of plant-related innovations, has probably been the most disputed issue of crop development over the past decades. IPR are expected to encourage private organizations to invest in plant breeding research so that new varieties are developed to the benefit of agriculture. Many organizations in developing and OECD countries, however, strongly oppose plant-related IPR protection for ethical or cultural reasons. They also disapprove of IPR on the grounds that it primarily serves the interests of the northern crop development industry. IPR protection is often regarded as a new form of protectionism, through which national industries are protected abroad without violating neo-liberal policies. The sharp controversy demonstrates the confusion over the precise benefits and disadvantages of IPR for agriculture in developing countries.

The intention of this book is to examine the socio-economic causes underlying the international controversy over the exploitation of genetic information. This is not a matter purely of academic relevance. Every attempt to achieve more international consensus on the way plant genetic information is to be exploited requires an accurate analysis of the relevant power constellation of the actors involved. In our view, such an analysis must include:

(a) an assessment of the exploitation of genetic information in relation to the production sector from which the information derives its value, rather than an assessment in relation to the legal context of international agreements on biodiversity and intellectual property protection;

(b) a thorough examination of the role and interests of the various actors involved in the controversy, such as governments, industry, consumers, non-governmental organizations, farmers, indigenous peoples, etc. It is often assumed that the conflicts are *inter*national and North-South in nature, while the *trans*national dimensions of the conflict are hardly considered.

The focus in this book is agro-food production. It examines the causes of the international controversy over plant genetic information as far as genes are used for the development of new plant varieties. Two main antagonistic positions can be distinguished. The first claims that genetic information that is derived from landraces or wild plants is just 'raw', a natural resource, hardly representing any economic value. Value is added by professional breeders and geneticists, who are able to detect, define, isolate, and use the specific information contained in the resources to develop new crops. No compensation is therefore required for the source person, community, or country from which the raw material was first obtained. The second, opposing position holds that genetic information *does* represent economic value in itself, irrespective of its further application. Consequently, the end-user of the genetic information has an obligation to compensate the original source person or source country. Scientists who freely take biological organisms out of developing countries are depriving these countries of natural resources with an economic value.

This perception of the controversy over genetic information as a matter of compensation has induced a growing number of non-governmental organizations and scholars to calculate the commercial value of plant genetic information. Whether such calculations are useful is doubtful, however, because the opposition to the free collection of plants and IPR protection in developing countries is not just based on a desire to obtain a share in industrial benefits.

In our view, the controversy over the exploitation of plant genetic information is much more than a distribution problem; it results from a complex combination of socio-economic differences, political relations, technological advancements and juridical changes. This assumption was confirmed during the research we carried out in Chile and Colombia. For example, in 1996 we had a meeting with representatives of the *Sociedad Mapuche Newen* in Temuco, 500 km south of Santiago de Chile. The Mapuche Newen, an indigenous community, cultivate potato varieties of which the genetic characteristics are frequently used by potato breeders around the world. The breeders have often been accused of biopiracy, since the indigenous people have never received any compensation for the genetic information contained in the landraces that have been collected. This situation aroused much anger among the Mapuche, and confirmed their idea that foreigners, including Chileans with Spanish roots, turn to them only to capture resources. However, the perceived injustice did not stand alone but was intrinsically related to daily, and more pressing problems which we were informed of extensively, including discrimination, violation of their ethnic and territorial rights, poverty, and unemployment. This reaction indicated that it is not just the remuneration of lost property that is at stake, but rather the general socio-economic marginalization to which peasant and indigenous communities in developing countries are subjected. Hence, the roots of the conflict over the exploitation of plant genetic information should first be sought in social, economic, and cultural circumstances.

Our approach is therefore interdisciplinary in character, located at the crossroads of science, law and political science. In a nutshell, our argument is that agro-food production can be organized in various manners, each manner bringing along a specific view on the function and the design of plant varieties and consequently also on their resources and exploitation rights. Thus, conflicting views on how agro-food production should be organized provoke conflicting perceptions of how genetic resources should be handled.

Before we map out the basic structure of the book it is necessary to explain two terms we use throughout: genetic information and crop development. Contrary to conventional usage in most of the literature on the present subject we prefer to speak of *plant genetic information* rather than 'plant genetic resources'. The term "genetic information" refers to the material of inheritance, DNA (*deoxyribonucleic acid*) consisting of *adenine, cytosine, guanine,* and *thymine* (Abercrombie *et al.* 1980:123,124). In this definition, a gene, a tissue, a plant, or even a whole population can be considered as a source, or carrier of genetic information. The term is deliberately chosen since 'plant genetic resources', normally defined as all living material "with an actual or potential value" (IBPGR, 1991:74; UNEP, 1992), has the disadvantage that it inherently carries a political meaning, because it presupposes that genetic information has economic value.

The second term, *crop development*, comprises all research and other activities that aim to alter the performance of crop plants. These activities may be directed at changes in the genetics of plants by means of, for example, selection, cross breeding and genetic engineering, or may focus on external inputs that influence the growth of plants. We prefer the term crop development over the conventional terms such as 'plant breeding', or 'crop improvement'; the first has a much narrower connotation than crop development and usually refers to crossing, while the second has the disadvantage that it positively colours the products of the techniques. A newly released variety is always denoted as improved by its creator. By using the term crop development, we underline the fact that crop development strategies may lead to varieties with qualities that may be praised by one and challenged by another.

Crop development is thus not a neutral, technological activity, but a subject of political decision-making in respect of agricultural production. Private and public plant breeders continuously deal with questions such as: what crops should be focused on; where and to what extent should new plant varieties be marketed; which genetic information should be conserved; how and to what extent should the resulting varieties be protected against reproduction by competitors. Answers to these questions depend on the strategies employed by farmers, public breeding institutes, private breeding enterprises, and governments to adjust crop development to a preferred type of agro-food production. In the course of this century, these strategies have evolved into 'crop development policies' on the use, conservation and protection of plant genetic information.

The complexity of the research subject has induced us to design an analytical framework that explores both the historical and spatial variation of agricultural production. In order to explain *historical variation* we go back to the emergence of the crop development industry in the early 20th century in the USA and several Western European countries. The more it was realized that agricultural competitiveness could be influenced by crop development, the louder the call became for specific governmental policies in this field. In most OECD countries, plant conservation and IPR protection for plant-related innovations have gradually become a prerequisite for further crop development. In developing countries, where the 'success' of industrializing agriculture varies greatly, the call for new crop development regulation is much less unanimous. This point illustrates the need for *spatial vari-*

ation in our analysis. Where little industrialization of agricultural production has taken place, the interest in adopting new conservation and IPR policies is low. Differences in the degree of agro-industrialization occur among countries, but also within countries, especially in the developing world. The assessment of the conservation and IPR policies differs accordingly.

In order to collect original data on conservation and IPR policies in developing countries, we carried out research in two Latin American countries, Chile and Colombia, in 1994 and 1996. We selected these two countries for a case study because both of them maintain a considerable degree of biological diversity within their borders. Colombia is even regarded as a country of 'mega-biodiversity'. Its tropical forests harbour a wealth of plant species with actual or potential commercial value, such as passionfruit, papaya, and numerous ornamental plants. Chile is rich in strawberry and potato species. In both countries this asset has turned the implementation of new conservation policies into a political event with both national and international dimensions. Moreover, Chile and Colombia are among the first developing countries adopting legislation for plant-related IPR protection, which in both countries has triggered considerable public debate.

The first chapter provides an overview of the controversy over the exploitation of plant genetic information. The most common analytical approach for explaining the controversy is discussed, as well as our alternative approach.

Chapters 2, 3 and 4 have a strong historical dimension in that they explore how, throughout the past 100 years, crop development policies have gradually been anchored in the process of agricultural industrialization in Western Europe and the USA. Chapter 2 describes the early governmental initiatives in these countries to regulate crop development in the period 1870-1930. A relation is established between specific conservation and IPR regulation and the specific position of the countries in the international division of labour. Chapter 3 is entirely dedicated to the 1930s. It was in this decade that two countries, Germany and the USA, were able to establish a close link between crop development policies and long-term strategies for agriculture. Chapter 4 examines how, after the Second World War, the USA managed to 'export' its crop development policy to Western Europe, and later, under the flag of the Green Revolution, to developing countries. Special attention is given to the role of the FAO in this process.

Chapters 5 and 6 deal with the present period, from the 1980s onwards. In the fifth chapter, the focus is on the latest stage in the process of agricultural industrialization: the emergence of the transnational biotechnology industry and the growing demand for global conservation and IPR regulation. The final chapter is devoted to the question of how agricultural producers in Chile and Colombia respond to new global conservation and IPR regulation. It is argued that the root of the conflicts should be sought in the frictions between those who support an industrialization of agriculture and those who do not.

Chapter 1

Crop Development and Agriculture

Controversies on plant genetic information first came into the public spotlight in the early 1980s. Not that they were a new phenomenon; ever since human beings were first able to design crop plants, there have been divergent views on the methods and objectives of crop development. For example, the production of hybrid maize varieties was disputed in the USA in the 1920s because the yields were poor and because they eliminated the opportunities of farmers to produce their own seeds. Some decades earlier, American and French fruit breeders had encountered strong opposition when they started to claim patent protection for their new varieties. And the necessity to conserve plants from around the world as a 'resource' for plant breeding was acknowledged as early as the 1930s. Worldwide collection was controversial, however, and would even result in the elimination of one of the greatest geneticists of the century, Soviet citizen Vavilov. Controversies on crop development in the 1980s differed from preceding ones in that they were no longer the exclusive domain of scientific experts. A broad public became interested in plant patenting, genetically engineered food, the degradation of biological diversity, and plant conservation. Moreover, because all these issues were placed into a North-South context, crop development disputes gained a new, international dimension.

This chapter demonstrates our search for analytical tools to explain contemporary conflicts on the conservation and exploitation of 'plant genetic resources' (PGR). Our analytical framework is built on three arguments. The first is that crop development is strongly related to agriculture. New plant varieties are created with a specific objective. They have to fit into the type of agricultural production that is envisaged. For example, a highly industrialized agricultural sector requires other varieties than the traditional farming sector. The second argument is that every country follows its own strategy to improve its national agricultural performance. The availability of natural resources, the degree of agro-industrialization, as well as the overall economic and political situation may differ widely across countries, and even within countries. This variation induces an international division of labour in agriculture and forces every country to adjust its goals for agricultural development to its specific position in this division of labour. The final argument is that the international variation in agricultural production also has a historical dimension. Agro-food production capacity has evolved over time, but its pace has varied among different countries. Individual countries have historically had different positions in the international division of labour in agriculture. The policies in support of agricultural development have changed accordingly.

These arguments indicate the scope of our analysis, which is strongly determined by dimensions of time and space. Crop development has always been at the service of agriculture. Since the requirements of agro-food production have differed both internationally and historically, the conditions for crop development have differed as well. It is in these dissimilarities that the root of the conflicts on the exploitation of PGR can be found, and this is the thread not only of this chapter, but also of the entire book.

In section 1.1, the history and main controversial aspects of the present controversy on the conservation and exploitation of PGR is summarized. In section 1.2, we discuss the disadvantages of the scarce literature that has analysed the PGR conflict from a theoretical point of view. Then, in 1.3 and 1.4, our analytical framework is constructed and explained.

1.1 The plant genetic resources (PGR) conflict

Public concern about crop development was first raised by a group of American, Canadian and European activists and researchers that gathered in Canada in 1977, under the aegis of the International Coalition for Development Action (ICDA), to discuss the world's food supply. Seed was a major issue, arising from the concern that "the genetic base of the world's food supply was quickly disappearing and that restrictive legislation was making it possible for agribusiness to gain control of this vital segment of the total food system" (Harmston, 1979). Seeds of the Earth was the title of the book that resulted from this gathering (Mooney, 1979). It was widely publicized and aroused a considerable reaction in developing countries and within the seed industry.

The concern for the effects of the prevailing crop development policies outlined in Seeds of the Earth was elaborated on by participants of the ICDA gathering and other authors in the 1980s (*cf.* Berlan, 1983; Mooney, 1983; Buttel, 1984; Buttel *et al.*, 1985; Doyle, 1985; Hobbelink, 1987; Junne, 1987; Fowler *et al.*, 1988; Kloppenburg, 1988a,b; Juma, 1989; Kloppenburg and Kleinman, 1989; Ruivenkamp, 1989). We summarize the arguments below because of their historical relevance. The publications reflected a new tendency in which civil pressure groups, in OECD as well as in developing countries, question the benefits of the accelerating industrialization of agriculture in general and of crop development in particular. The main arguments put forward were the following.

(a) The prevailing crop development policies are destroying the genetic base of plant breeding

It was pointed out that most crop varieties used in agriculture in the 'North' or OECD member countries, do not originate from the temperate zone, but are derived from plants that were introduced from the 'South', or developing countries. This implies that plant breeders in the North must return to the genetically more diverse South for genetic resources. 'Fresh genes' are necessary to improve crop yields, to increase pest or insect resistance, or to adjust varieties to adverse ecological circumstances. However, due to the rationalization of agriculture, the genetic diversity is gradually being wiped out. Especially the spread of new and uniform, high-yielding varieties during the Green Revolution accelerated the replacement of landraces (the traditional farmers' varieties) and wild relatives of crop plants. In other words, it was contended that the industrialization of agriculture greatly facilitated a process of 'genetic erosion' in the South, which would lead to the destruction of the basic resources for future agricultural production.

(b) Crop development is based on a South-to-North 'gene drain'

In most publications it was held that, for centuries, colonial powers and later Northern industrialized countries had freely collected seeds and plants of landraces and wild relatives in the South. The seeds and plants collected were transferred to the North where they were stored in botanical gardens or, more recently, in so-called 'genebanks'[1], beyond the control of the countries of origin. As landraces and wild relatives were considered to be a public resource - a 'common heritage of mankind' - property and payment questions were never raised. The authors argued, however, that landraces and wild relatives represent considerable economic value. Constituting the basis for most agricultural crops, genes derived from the seeds and plants collected in the South would have added billions of dollars to the economies of industrialized countries. Moreover, advances in genetic engineering have expanded the number of plants with potential value for plant breeding and also pharmaceutical production. The countries of origin, it was pointed out, have never been remunerated for this 'gene-drain'.

(c) Multinational enterprises take control of the seed industry

Of major concern in the publications were the effects of the privatization of crop development. Since the 1970s, agro-chemical, pharmaceutical, and food processing corporations had become interested in the seed sector and were taking over the family-based seed firms. It was argued that the increasing involvement of private capital in crop development would lead to a situation in which a few multinational enterprises would gain control of the world's food production. The new seed companies would privately hold unique seed collections, breed varieties which fit their chemical, mechanical and processing interests (but which are not necessarily beneficial to farmers and consumers), produce plants that are genetically uniform (thus affecting genetic diversity) and cannot be reproduced, and protect their new varieties with intellectual property rights.

(d) Intellectual property rights hinder transfer of crop development technology to developing countries

Finally, these publications pointed out that plant breeding and the propagation and marketing of new plant varieties, was being hindered in developing countries, because of intellectual property rights granted to plant breeders on their new creations. Thus, even though the new varieties contained genetic information freely obtained from developing countries, unauthorized exploitation of these varieties has become increasingly difficult. This situation was considered as unfair and disadvantageous for agricultural production in developing countries.

In the early 1980s, the 'Group of 77' developing countries began to raise questions about the exploitation of genetic information. The issue fitted the struggle of developing countries for a New International Economic Order (NIEO). The NIEO was an attempt by developing countries to distribute global wealth better, and was adopted as United Nations (UN) policy in 1974. The developing countries used their national sovereignty and their numeric majority in UN forums to challenge political and economic activities - of transnational enterprises in particular - that were sustaining the unequal distribution. Both the 'gene-drain' and the legal restrictions in the North-to-South transfer of new plant varieties were new illustrations of general trends which developing countries had addressed since the early 1960s. It was especially the 'gene-drain' issue that transformed the FAO deliberations in the period 1981-1983 into fierce North-South confrontations on PGR. The term resources emphasizes the economic and political value attributed to genes.[2] These conflicts intensified because most of the world's seedbanks, the storage facilities for the collected seeds, plants and plant tissue, were under control of OECD countries. This implied that developing country governments could not exert control over the genetic information contained in the stored plant material. The issue became all the more important when it was revealed that governments of OECD countries had occasionally embargoed the release of seeds stored in their territory to hostile governments (*cf.* RAFI, 1988).

The international conflict over the exploitation of PGR (in short: the PGR conflict) began to take shape during the controversy over two FAO resolutions which were at stake in the period 1981-1989, concerning the International Undertaking on Plant Genetic Resources, and the establishment of an FAO Commission on Plant Genetic Resources (CPGR). The resolutions had been proposed by a core group of developing countries, which were supported by Northern non-governmental organizations (NGOs) affiliated to ICDA. Both resolutions were adopted in the FAO in 1983, despite opposition from most OECD countries.

The International Undertaking was a non-binding agreement that provided standards and rules for the conservation and exchange of seeds, plants and plant tissue between countries. The Undertaking also formalized the *de facto* status of PGR as a 'common heritage of mankind'. At first glance this formalization seems in line with OECD country preferences, but it was not. The developing countries had used their majority to adopt a very broad definition of PGR. As a result, the Undertaking considered not only wild plants and landraces, but also commercial plant varieties - legally protected or not - as a 'common heritage' and thus publicly available for breeders and farmers worldwide.

A first step in resolving the controversy was taken in the FAO in 1989. In that year most developing and OECD countries achieved agreement on three principles. First, it was accepted that plants protected under the system of plant variety rights were not freely available. Thus, developing countries were prepared to accept and recognize intellectual property rights on plant varieties. Second, it was agreed that 'free access' to landraces and wild relatives did not mean 'access free of charge'. This implied that public and private plant breeders would consider payment for the plants and seeds they collected in developing country territory. Third, a new type of rights, referred to as 'Farmers' Rights', was adopted. These rights were not defined but only justified by referring to the farmers' efforts for thousands of years in domesticating plants.

The Farmers' Rights concept had originally been proposed in 1985 by a number of farmer-oriented NGOs led by the Rural Advancement Foundation International (RAFI), a Canadian NGO, and ICDA. With its concept, the NGOs were able to link their concern for further genetic erosion with the contentious issue of 'gene drain'. Farmers' Rights can probably best be understood as *collective* intellectual property rights which entitle farmers to receive financial support from an International PGR Fund governed by the FAO. However, the contributions to the Fund were voluntary for participating countries, so that neither the International Fund nor the Farmers' Rights concept have materialized.

In later years, the course of the PGR conflict was greatly influenced by two other international agreements: (a) the Convention on Biological Diversity (CBD), adopted during the United Nations Conference on Environment and Development (UNCED) in Rio de Janeiro in 1992, and (b) the agreement on Trade-Related Aspects of Intellectual Property Rights (TRIPS), which was part of the new General Agreement on Tariffs and Trade (GATT), concluded in Marrakesh in 1994.

The CBD was the outcome of an international negotiation process that originated in public environmental concern in OECD countries. Rather than being focused on plants useful for agriculture, as is the FAO's Undertaking, the CBD is directed at the preservation of all plants and other organisms sustaining the global eco-system. The most contentious issues in the PGR debates in the FAO also emerged during the CBD preparations: the North-South divide on the distribution of benefits from biological organisms, the property rights on these organisms, and the access to technology required to exploit biological organisms. However, the solution to the controversies in the CBD differed considerably from what had been agreed in the FAO Undertaking. The CBD, on the one hand, recognized that many countries had already implemented intellectual property protection of technology involving biological material. On the other hand, the Convention eliminated the 'common heritage of mankind' status of wild plants and landraces. Instead, it was accepted that 'countries of origin' had sovereign rights over all biological material (plants, animals, micro-organisms) originating in their national territory. The CBD thus gave plants the status of national property, providing a legal basis for 'benefit sharing' arrangements.[3] The CBD also recognized the rights of indigenous and local communities to share benefits.

The second agreement, that on TRIPS, was the result of an initiative of the world's main industrial interest organizations to protect new technology, medicines and audio-visual works better against imitation. Legal protection of innovations related to biological material was initially not so relevant, but emerged as a negotiation topic around 1990, four years after the start of the Uruguay Round (Van Wijk, 1990). By that time, the first inventions in biotechnology had been protected under patent law and had found their way onto the market. The large chemical and pharmaceutical companies that had become involved in genetic engineering and crop development began to realize that worldwide legal protection of their biotechnological innovations would soon become essential in order to defend their leading edge. A group of developing countries opposed the plans to strengthen the international patent system. They advocated the exclusion from patentability of (among other things) plant or animal varieties and other products or processes if this was required on grounds of public interest (GATT, 1990). Despite this opposition, the TRIPS agreement was signed by 125 countries in 1994 as part of the new GATT. The agreement implies that legal protection of crop plants is recognized in most parts of the world.

1.2 'International Regimes' become comfortable objects of analysis

Since the FAO ceased to be the dominant forum for negotiating the exploitation of plant genetic information, the PGR conflict has no longer been tied exclusively to the production sector that depends on plant genetic information: agriculture. Instead, the PGR conflict spilled over into negotiations on the CBD and TRIPS agreement. As part of the CBD, the questions about the access to, and conservation of landraces and wild plants are now related to the broader issue of biodiversity preservation for the sake of 'sustainable development'. Under the TRIPS agreement, the exploitation of new plant varieties is also generalized and considered as the exploitation of a protectable 'innovation'. This common denominator gives plant varieties equality with, for example, computer chips, movies and compact disks.

The fragmentation of the PGR conflict over three international treaties has induced a growing body of literature in which the conflict is separated from its political-economic context (*cf.* Hamilton, 1993; Rose, 1993; Sánchez and Juma, 1994; Leskien and Flitner, 1997; Mugabe *et al.*, 1997). Rather than taking into account the conflicting interests in agriculture, these publications tend to reduce the PGR conflict to a series of tough negotiations at international forums between Northern and Southern states on legal provisions that regulate a 'sharing of benefits' derived from genetic information.

In more analytical contributions concerning the PGR conflict, the strict focus on regulatory disputes seems to gain in popularity. This is largely due to the preference among social scientists for the 'Regime theory', which focuses on the international regimes under which states interact in order to solve an international problem. Although the definition of international regimes was first formulated more than 20 years ago by Ruggie (1975:570), the question of what precisely are regimes has remained a matter of dispute. Most generally accepted is the definition of Stephen Krasner, who characterizes regimes as "principles, norms, rules, and decision-making procedures around which actor expectations converge in a given issue area." Krasner refers to 'principles' as coherent sets of theoretical statements about "how the world works". 'Norms' specify standards for behaviour, while rules and decision-making procedures refer to specific prescriptions for behaviour in clearly defined areas (Krasner, 1982:186).

Regime theorists often present principles, norms, rules and decision-making procedures as 'intervening variables' between the political and economic power of states and their actual political behaviour (*cf.* Keohane and Nye, 1977; Krasner, 1982; Rittberger, 1988). States involved in a regime tend to negotiate with each other, or at least listen to each other's positions. As a result, narrow national interest calculations are moderated so that the number and intensity of international conflicts can be reduced. 'Strong states' participate in a regime with 'weak states' and accept compromises to maintain power at a minimal cost. Cooperation with weaker states often involves less political risk (and is cheaper) than confrontation and full (military) control. Weaker states, on the other hand, may also benefit from a regime, as it allows them to avoid confrontation with other, more powerful states. These mutual benefits create the basis of 'collective action' in international relations.

Regime analysis has been the major social science approach to study negotiations on the 'global commons' within the United Nations (*cf.* Krasner, 1985; Young, 1989; Wolf, 1991). The global commons (such as the high seas, space, and the polar regions) are among the issues which strong states prefer to manage through collective action, rather than unilateral control. This strategy should prevent a 'tragedy of the commons'[4] and ensure that exploitation benefits all people. When the Food and Agriculture Organization of the United Nations (FAO) in 1983 officially declared plant genetic resources to be part of the 'common heritage of mankind', the regime analysis was the most obvious theoretical approach to examine the conflicts on the management of this new global common.

We discuss three recent regime contributions on biodiversity and intellectual property rights negotiations in relation to PGR. The first is from Arts (1998), who places the CBD within what he calls the "nature conservation regime", but does not tie this regime to a specific organization or treaty. Nor does he interpret regimes as the sum of explicit principles, norms, etc., formalized in treaties and organizations. In his conception, regimes are marked out less visibly. They consist of both formal and informal rules which "frame (common) actions of state and non-state actors" in a specific issue-area. The second contribution, from Spector *et al.* (1994), analyses the UNCED which is presented as a step towards an "umbrella regime for sustainable development" which may spawn sub-regimes in particular environmental issues, including biodiversity.

The publications of Witmeyer (1994a and 1994b), the third regime contribution, focus on PGR for agriculture in particular. The author's concept of a regime is much narrower than that of Arts and Spector *et al.*. Witmeyer identifies three regimes on the basis of three international agreements: the "PGR regime" based on the FAO Undertaking; the "biodiversity regime" based on the CBD; and the "intellectual property rights (IPR) regime" founded by the agreements on TRIPS and UPOV (the International Union for the Protection of new Varieties of Plants). According to the author, the conflicts within the PGR regime are a "collective action problem" which result from the limited financial and political commitment on the part of industrialized countries. The weak legal and financial basis of the Undertaking has handicapped the implementation of its main principle: the 'common heritage of mankind' in respect of PGR. The younger biodiversity regime promotes a more market-driven regime for the conservation of biodiversity, including PGR of agricultural crops. Being a legally binding contractual agreement, the rules of the CBD have legal precedence over the FAO Undertaking. Witmeyer argues that biodiversity conservation based on economic incentives, as promoted in the CBD, may harm the free availability of PGR, which he considers as particularly relevant to public plant breeding in developing countries. Moreover, he doubts whether the CBD will result in large revenues for source countries because of the renewable and 'non-rival' quality of the wild relatives and landraces of most food crops.

Despite the wide differences in applying Regime theory, the three publications have a number of characteristics in common. Notably, they all approach international conflicts from a managerial perspective. The basic question underlying the books is *how* common principles, norms, and procedures in the area of nature conservation and nature exploitation can be established so that collective action is possible. Questions as to who is to gain from the regime and who is to lose are hardly considered.

This limited scope of analysis becomes clear in the way the 'issue-areas' are approached. None of the books provides a framework in which the utility of genetic information is examined in such a way that the relevance of its exploitation for distinct producers and consumers is explained. Such an account would necessitate an examination of the political strategies involved in the regime formation. Instead, the authors tend to follow the negotiation agendas prepared by the respective secretariats, and even use the distinct articles of the international agreements at issue as the thread of their analysis. By reducing the PGR conflict to a dispute over legal articles that are being negotiated in at least three different international forums (FAO, UNCED, GATT/WTO), the regime approach inevitably results in a fragmented, administrative and legalistic perception of the conflict at stake.

That the three publications are preoccupied with the question as to how a regime develops, rather than who is to benefit from it, is exemplified in the approach of non-state actors involved in regime formation. In the negotiation processes on the International Undertaking and particularly the CBD, a large number of NGOs have been officially involved, which has made them unavoidable research objects.[5] The attention devoted to these NGOs, however, is induced by only one characteristic they have in common, namely that they are non-state actors participating in international regimes. This managerial focus results in a rather technical analysis of the NGOs' role in international regime formation. Witmeyer pays attention to some NGOs as 'critics' of the PGR regime, but it is not at all clear to what extent and in what direction they have influenced the FAO Undertaking. Spector *et al.* devote an entire chapter to the role of NGOs during the UNCED and conclude that many NGOs are disappointed about their impact on the official UNCED process. The editors, however, state with tacit satisfaction that the NGOs have been useful in assisting national delegations as well as in opinion-building in the various countries, while they also may be helpful in implementing the agreements (Spector *et al.*, 1994:234,238). Arts, for his part, meticulously measures the influence of international NGOs on the final drafting process of the CBD. Article by article, he judges the impact of NGOs in terms of their ability to win national negotiators for their point of view.

Left out of these three publications is the political dimension. NGOs represent business interests or social movements. They have political and cultural principles and economic interests that may greatly diverge. Even though in one of the contributions it is acknowledged that regime formation does not take place in isolation but is influenced by international political and economic forces (Arts, 1998:71), the NGOs are not related to them.

Neither is the authors' conception of government helpful in providing an insight into the political structure of the conflicts. All three authors consider governments as if they were monolithic units that take unequivocal positions in international relations. Obviously, governments are forced to present one more or less consistent position when it comes to the adoption of an agreement. Difficulties certainly arise when states adhere to different international treaties with conflicting obligations. Even though this formal position is a political reality, it would be an error to take it as the only point of reference in political analysis. The formal position reflects the dominant economic and political strategy within a government. Alternative positions, however, exist and continuously play a role during regime formation, implementation, enforcement and amendment.

Particularly in respect of the PGR conflict, many governments have been internally divided. Nevertheless, neither Witmeyer, nor Arts take into account that the government is a many-headed organism, which may take different or even contradictory positions at different international forums. For example, the Ministries of Agriculture in OECD countries have been representing their countries in UPOV, where they advanced plant variety protection as a legal protection system for plants. The Ministries of Industry and of Trade of the same countries, however, considered patent protection for plants during the Uruguay Round of GATT negotiations. At the same time, the Ministries of Environment and of Development Cooperation, for their part, tended to play down the importance of IPR for plants, or even criticized such protection during the UNCED process. Another example is that Ministries of Environment tend to stress the global importance of preserving plant species as carriers of large ecosystems. The Ministries of Agriculture, however, often prefer to use the limited conservation funds for 'agro-biodiversity' (i.e. genetic resources of crop plants), while the Finance Ministries and those of Industry and Trade have never been enthusiastic about additional public funds for plant conservation. Such a diversity of attitudes within the government is only recognized in the book of Spector *et al.*, but has subsequently been considered a management problem, hindering regime formation.

Overall, the three regime analyses do offer detailed and interesting overviews of the international decision-making processes on crop development policy. As political analysis, however, they are of limited value, as they tend to reduce the PGR conflict to a series of negotiations between (Northern and Southern) governments about legal articles and financial resources. From such analysis it is hardly possible to reconstruct the political and economic interests involved and the extent to which these interests are reflected in the regimes that are negotiated. Moreover, the analyses cannot provide any insight into the increasingly transnational character of the conflict. The North-South alliances between governmental departments, business organizations and non-profit NGOs on the basis of common views as to how PGR should be exploited cannot be discovered and explained through regime analysis. In our opinion, however, contradicting views between such transnational alliances are at the heart of the PGR conflict.

1.3 Analytical framework of the book

Considering the limited value of the Regime theory in explaining the international PGR conflict, we propose an alternative analytical framework which relates the PGR controversy to changes in agro-food production. The international conflicts emerged when crop development entered a period of major transformation. Since the mid-1970s, massive capital investment in genetic engineering has accelerated a process in which crop development is genuinely being industrialized. This process changed the value attached to plant genetic information, reshuffled the power constellation of actors involved in crop development, and rearranged governmental intervention.

The transformation of crop development is not an isolated event, however. It is part of an overall process of industrialization of agriculture, which began in the 19th century or earlier and has not yet come to an end. The socio-economic transformation that has been related to this process particularly provoked resistance among those parts of society that could not benefit from the changes. We consider the opposition to the replacement of landraces by new, uniform varieties, the 'gene-drain', the resistance to plant patenting and to the increasing importance of transnational seed companies, to be a reaction against the overall industrialization, not only of crop development, but also of agriculture in general.

In the following section, the PGR conflict is placed within the context of changing conditions for agricultural production. We first examine the industrialization of agriculture as such, and then the opportunities to support, steer, or resist this industrialization process, or support non-industrial improvement of agriculture. In the final section, we focus on the historical and international differences in agro-industrialization.

1.3.1 Industrialization of agriculture

In political economy, the agricultural sector traditionally has been considered an anomaly in capitalism. In the 19th century, all production sectors attracted capital investment, which induced a replacement of the traditional craft work by mechanized production processes and wage labour. Agriculture remained the exception, however. Agrarian capitalism did develop in Europe and the USA, but ended up in a "miscarriage" around the 1870s, basically because the capitalist farm enterprise could not compete with the family farm (Koning, 1994). This raises the question why capital investors failed to organize a superior production system that undermined the competitive advantage of household production.

According to Goodman *et al.* (1987) the basic constraint for private capital investment in agriculture has been the inability to eliminate the risks and discontinuities intrinsic to a natural or biological production process. In their 'theory of agro-industrial development' they hold that, unlike the sectors of handicraft activity, agriculture could not be directly transformed into a branch of industrial production. There was no industrial alternative to the biological conversion of amino acids and solar energy into food. Agriculture relies on outdoor space, on climatic conditions and biological reproduction cycles, which are difficult to control.

Instead of removing the technical obstacles directly, the industrial transformation of agriculture has taken place through two long-term processes: "appropriation" and "substitution". *Appropriation* refers to a process in which the controllable elements of farming are being transformed into industrial activities and reintroduced into agriculture as inputs. Facilitated by scientific advances, ever more elements of the biological production process could be controlled and replaced by industrial activities and industrial labour. Major examples are the replacement of horses by tractors and of manure by synthetic fertilizer. Thus, industrial capital has been gradually 'appropriating', or industrializing agriculture. While appropriation tends to control the volatile and unpredictable processes of agriculture, it remains dependent on a rural-based production process. *Substitution*, however, tends to eliminate the material base of agriculture. Substitution refers to the process directed at industrial food production and tends to undermine the relationship between food and fibre production on the one hand, and the countryside on the other. Agricultural products are increasingly reduced to interchangeable industrial semiproducts (e.g. starch, or carbohydrates) and may eventually be substituted by synthetic products (e.g. artificial sweeteners) (Goodman *et al.*, 1987).

The concept of appropriation is very suitable for explaining the historic transformation of crop development. This process began in the early 20th century in

Europe and the USA, and involves the gradual replacement of landraces by scientifically bred commercial plant varieties. Farmers increasingly buy fresh, commercial seeds instead of using their traditional varieties and on-farm saved seeds. The industrial appropriation of crop development is severely hampered by nature, however, as plants do reproduce themselves. This natural mechanism enables farmers to re-use seeds in principle indefinitely and significantly lowers the demand for commercial seeds. Thus, in order to make crop development attractive for private investment, nature had to be controlled.

Kloppenburg (1988a) has extensively described the subjection of crop development to market economy principles, a process he dubbed the "commodification of seed". Kloppenburg's starting point is the dual character of seed. When seed transforms into a plant, the original seed is in fact replaced by manifold new seeds, which may be used as product (as grain), or may be employed as a means of production (as seed) for planting the next crop. The duality of seed, which enables its use for growing a crop and for producing seed, has been fundamental to the expansion of agriculture. Farmers always saved a part of their crop as seed, for replanting purposes or for exchange with other farmers. They used to purchase seed of a new variety only once, because they could reproduce the seed by themselves in the following years. This practice, however, hindered agricultural entrepreneurs who tried to make a living out of crop development. On-farm saved seed is cheaper than commercial seed and, overall, exerts a downward pressure on seed prices. Thus, the elimination of the dual use of seed was a condition for maximizing the expansion of the commercial seed business. This could only be achieved with active support of the state. In his book, Kloppenburg maps out the crop development policies of the U.S. government in the 1920s and 1930s, which laid the foundation for the process of seed commodification. Public breeders developed so-called 'hybrid' varieties that were deliberately designed to obstruct the biological reproduction process.[6] For biological reasons, only the first generation of such a hybrid performs well. The yields of the second and further generations of hybrids will decline and are generally less attractive to farmers. Thus, the dual character of seed is effectively eliminated by hybrids. They produce grain or fruit rather than useful seed. Farmers will shift from landraces to commercial hybrid varieties, if the hybrids significantly out-perform the traditional varieties. This knowledge was the incentive particularly for the U.S. seed industry to invest in hybrids as soon as the technology became available in the 1930s. Since, for technical reasons, hybrids could not be produced in some important crops, such as wheat and soyabean, the dual-use value of seed was also tackled by legal means: intellectual property rights. Through the grant of a patent or a plant breeders' right certificate, breeders are entitled to prevent or restrict unwanted propagation of varieties they have developed.

Whilst it is convincing to present the dual character of seed as a major obstacle to capital investment in crop development, the argument, in our view, is incomplete. The biological make-up of plants not only enables farmers to 'copy' and multiply the genetic information contained in the plant by saving seed, it also facilitates competing breeders to make use of that information. By crossing his own variety with that of a competitor, a plant breeder can freely acquire genetic information encoding for commercially valuable plant characteristics. In fact, the free exchange and crossing of plant varieties that have valuable traits has been of fundamental importance to agriculture. Farmers and breeders have built on each other's work so that yields have increased and quality has improved. However, the free access to the genetic information of a plant and the free exploitation of that information has also been an obstacle to capital investment. Controlling the transfer of valuable plant characteristics into the varieties of competing breeders was almost impossible. For this reason, we maintain that it is not the *dual character of seeds*, but the *triple character of plant varieties* that constitutes the major obstacle to capital investment in crop development. By nature, plants can be used for growing a crop, for propagation, *and* for creating a new plant variety. The 'appropriation' of crop development therefore requires the elimination of the latter two opportunities.

Ever since plant breeding commenced, the free use of varieties for breeding purposes has been a sensitive issue, but it became a real problem only in the 1970s. The free access to genetic information by breeders began to be questioned particularly when genetic engineering and promising commercial prospects in agriculture raised the interest of large industrial capital in crop development.

1.3.2 Rival agricultural production strategies

Although the theory of agro-industrial development is helpful in analysing the dynamic forces in crop development, it also raises questions. A major question involves the role of the state in agro-industrialization. Why have governments been facilitating processes of appropriation and substitution?

In their respective contributions, Goodman *et al.* (1987) and Kloppenburg (1988a) assume that strong governmental support has been a prerequisite for the industrialization of agriculture in general and the commodification of seed in particular. The authors, however, tend to simplify the concept of the state as such, as well as the political processes that determine state intervention.

Goodman *et al.* (1987) consider the state as an institution with a specific, and apparently pre-determined "task" to serve the interest of capitalist development. It is argued that, in ancient civilizations, the state was involved in food procurement and storage. During a "second major phase of state intervention" the state designed policies to reform feudal land structure, while in a third phase, emerging with the advent of industrial appropriation, state intervention has been in support of industrialization of agriculture. This is a rather teleological conception of the state. Even though the state indeed has played a pivotal role in the industrialization of agriculture, one should take into account the fact that state intervention is determined by political struggle, the outcome of which is, in principle, contingent.

Kloppenburg (1988a) demonstrates more ambivalence in his conception of the state. On the one hand, he implicitly recognizes that the role of the state in the industrialization of agriculture is not predetermined. He shows that public agricultural research policies and the intellectual property rights systems in the USA were changed in favour of the seed industry, only in response to an effective lobby of private seed companies. On the other hand, Kloppenburg follows the North-South dichotomy and tends to generalize the role of the state within this divide. The historical role of the U.S. government as facilitator of the industrialization of seed production is projected onto all core capitalist countries. He overlooks the fact that the private seed industry in Europe has never had the political impact of its American counterpart. As regards developing countries, Kloppenburg only presents one voice from the 'South', the voice of opposition to seed commodification. By doing so, he assumes complete consensus within states, between states, and between states and non-state actors in developing countries. The wide diversity of interests and strategies regarding crop development across the 'South' is neglected.

Both Goodman *et al.* and Kloppenburg seem to ignore the fact that governments have more options for their involvement in agriculture - options that also have considerable relevance for the way a government intervenes in crop development.[7] In our opinion, the state may:

(a) advance a market-led industrialization of agriculture by facilitating the removal of technical and political obstacles to the 'appropriation' of agricultural elements. The intended rise in productivity lowers production costs and farm prices, makes exports more competitive, allows for lower wages, and is beneficial for the processing industry;

(b) actively steer the industrialization process on the basis of a national political agenda. Socially undesirable developments in agro-industry are slowed down in order to reduce the social costs, while other desirable developments are stimulated. For example, governments may protect those national farm sectors that are not internationally competitive, or they may restrict the application of genetic engineering for ethical reasons. On the other hand, governments may encourage organic farming as an alternative route for agro-industrialization;

(c) support non-industrial forms of agriculture. This option is particularly relevant in developing countries where the peasantry cannot be incorporated in the agro-industrialization process, because they lack the necessary land, capital, infrastructure and markets. Facilitating non-industrial development of agricultural production on the basis of locally available knowledge and natural resources is a policy to combat rural poverty and food shortage.

The three options of state intervention presented here constitute the basis for three alternative and conflicting strategies for the organization of agriculture. In this publication we distinguish a 'market-led' agro-industrialization strategy, a 'state-led' agro-industrialization strategy, and a third strategy directed to non-industrial, farmer-oriented agricultural production. We will refer to them as *rival agricultural production strategies*. These three strategies must be considered as ideal types.

They are constructions of agricultural production strategies of public and private organizations, in the most pure form. The strategies encompass all socio-economic, technology and legal policies that are intended to speed up, steer or slow down agro-industrialization, or support non-industrial agriculture. Which of the strategies will prevail over the other in a particular country depends on national and international political processes.

The typology of rival agricultural production strategies will be used as an analytical concept to help us better understand the differences of opinion about how agriculture should develop and, consequently, what crop development policy should be followed. We maintain that in the most advanced capitalist countries the political dispute in the past century has primarily been located between the market-led and the state-led production strategy. The third strategy, directed to the support for non-industrial agriculture, has become relevant with the introduction of the Green Revolution, when in developing countries a large part of the rural population was marginalized. Particularly the peasantry, including the indigenous communities, could not be incorporated in the process of agro-industrialization, while no alternative employment opportunities in other economic sectors were available.

We propose to locate the roots of the PGR conflict in the existence of rival production strategies. The value attached to the conservation of landraces and wild plants, the desirability of a free exchange of PGR, the perception of hybrids technology and of intellectual property protection, and the attitude towards transnational seed companies, is, in our view, determined by the preference for one of the three strategies. Thus, the PGR conflict involves not just crop development policy, but the strategy for organizing agricultural production in general.

1.3.3 Historical Agro-Food Orders

The rival agricultural production strategies concept is intended to explain political differences in the organization of agriculture within individual countries and among different countries. It is necessary however, to examine the reasons for these differences. Why do we find a political majority in favour of the 'market-led' agricultural production strategy in country X, while in country Y the 'state-led' strategy is dominant? In order to find an answer to this question, we must consider that the degree of agro-industrialization varies considerably among countries. It is this variation that induces an international division of labour in agriculture. The position of a country in the international division of labour, as well as the historical changes in that position, greatly influence the national political processes involving agricultural production strategies, including the associated crop development policies.

In their classic paper, Friedmann and McMichael (1989) have worked out the national and historical differences in agro-industrial production and consumption. They organize their argument around the concept of *food regime*, a concept that "links international relations of food production and consumption to forms of accumulation broadly distinguishing periods of capitalist transformation since

1870" (Friedmann and McMichael, 1989:95). The food regime concept is theoretically connected to the 'Regulation school'. This school of thought in political economy has developed a theory on "social regulation", an institutional framework with corresponding regulating norms, basically involving the wage relation, which organize capital accumulation during a specific period of time: the "accumulation regime" (*cf.* Aglietta, 1979 and 1982; Bonefield, 1988; De Vroey, 1984). Historical changes in social regulation would have prevented the capitalist system from collapse during periods of crisis. The Regulation school distinguishes two accumulation regimes. The first is placed in the period 1873-1914. The second accumulation regime had its heyday between 1945 and the 1970s and is usually referred to as "Fordism".

In order to analyse historical trends in international agricultural production and trade, Friedmann and McMichael distinguish two food regimes that correspond to the two accumulation regimes from the Regulation school. The 'first food regime' was based on the wheat and meat exports from the settler states (mainly the USA) to European metropolises. These exports facilitated the urban industrialization in Europe and provided the capital that enabled the import into the settler states of manufactured goods, essentially for railways. The 'second food regime' was based on the integration of the former colonies in the international division of labour as newly independent states. They began to import wheat, particularly from the USA, at the expense of domestic food production, while their traditional markets for tropical exports, such as sugar and vegetable oils, declined due to import substitution policies in OECD countries. At the same time, the emerging transnational agro-industry, which centred on meat and new processed food production chains, began to restructure and integrate agriculture worldwide (Friedmann and McMichael, 1989).

The advantage of the Friedmann and McMichael approach is that it adds historical and spatial specificity to the basic tendency of agro-industrialization. Nonetheless, the food regime concept also raises questions. The authors neither define the concept of 'regime' thoroughly nor do they explain its origin. This is particularly confusing because the food regime concept has also appeared in the context of Regime theory analysis, a different theoretical approach that was discussed above in section 1.2. This conception of food regime, defined as a set of principles, norms, and decision-making procedures which regulate and control the "global food system", was used in publications that explored the nature and severity of world food problems and the effects of the active international diplomacy of food affairs around 1980 (Hopkins and Puchala, 1978; Puchala and Hopkins, 1982). Friedmann and McMichael (1989), however, explicitly refer to the Regulation theory, but they do not explain *how* their conception of food regime fits into that theory.

The focus on the production level supports the assumption that the food regime is considered as a 'sub-species' of the accumulation regime used in the Regulation school. However, Aglietta and other regulation theorists deal especially with 'social regulation', while Friedmann and McMichael in their paper pay little attention to the regulatory system that has conducted the historical patterns of (inter)national relations in agriculture. Where they do mention regulatory elements (e.g. post-war food aid and agricultural support measures), their argument suffers from economic determinism. The role of the state (and of regulation) seems to be pre-determined by economic requirements, and they fail to take into account the political conflicts about the direction and aims of regulation in agriculture, as well as the unique national socio-cultural and techno-economic settings in which these conflicts take place.

Rather than food regime we propose a new, but related analytical concept: the *historical Agro-Food Order*[8]. The historical Agro-Food Order refers to a specific period in agricultural production that can be distinguished on the basis of (a) a particular international division of labour in agriculture, and (b) a regulatory framework that facilitates this temporary division of labour. The term "order" may sound at odds with a world which is typically chaotic and where social and (international) political relations are almost inherently volatile. We do use the term nevertheless, to indicate a period of relative stability of major tendencies in international agro-industrial trade and related social regulation.

The *international division of labour* in agriculture refers to the global production and distribution of farm products or industrially produced food. Countries have different positions in this international division of labour because of their dissimilar capacity to produce, export and import agricultural products. This capacity is related to the degree in which their agricultural sector has been industrialized. In individual countries a *regulatory framework* is designed in a political process by governmental departments, private industry, farmers' organizations and other non-governmental, non-profit organizations. Each of these actors may advocate one of the rival agricultural production strategies, discussed above. Depending on the economic and political strength of the actors involved, one of the agricultural production strategies will become dominant.

We distinguish between three historical Agro-Food Orders, basically following the food regime periodization of Friedmann and McMichael (1989). The First Agro-Food Order is situated in the period between the 1870s and the 1930s, the Second between the 1930s and the 1980s, while the Third Agro-Food Order began in the 1980s and will presumably have its heyday in the 21st century. Each of these Agro-Food Orders is thus characterized by a historical international division of labour in agriculture with a corresponding dominant agricultural production strategy. The latter commenced as a purely national strategy in the First Agro-Food Order, but has become increasingly international in the Third, due to international political integration and globalization of the economy.

We use the Agro-Food Order concept in finding an explanation for the PGR conflict. In each of the historical Agro-Food Orders, a dominant agricultural production strategy, including an associated crop development policy, can be distinguished. The relationships between the Order, the strategy and the policy have been visualized in Figure 1.1. For the sake of clarity, the figure exhibits only the internationally dominant agricultural production strategy and crop development policy in a given period, ignoring the variation among different countries and within individual countries. This variation, however, is discussed in more detail in later chapters.

Figure 1.1 Crop development policies in three historical Agro-Food Orders

Agro-Food Order	International division of labour in agriculture	Dominant agricultural production strategy	Aim and features of dominant crop development policy
FIRST 1870s-1930s	**Settler states, notably USA, export grain to feed urban labour in European metropolises** • Internal colonial trade of tropical crops loses significance vis-à-vis emerging international market for temperate agriculture.	**Initial state intervention to support national temperate agricultural sector in industrial core countries**	**Improving competitiveness of national temperate agricultural sector** • Focus of crop development shifts away from exchange of tropical plants among colonial territories to introduction of foreign landraces in temperate areas. • Emergence of national, public agricultural research system. • Early state control of seed market. • Development of hybrid technology in public maize breeding (USA). • Private industry attempts to patent ornamental and fruit varieties.
SECOND 1930s-1980s	**North America and Europe become food exporters. Developing countries emerge on world market as importers of temperate cereals** • Emergence of a transnational agro-industry centred on wheat, meat and processed foods.	**State-led industrialization of agriculture** • Deep state involvement in agriculture facilitates emergence of transnational agro-industrial capital. • State-led agro-industrialization exported to and adopted by developing countries in 'Green Revolution'.	**Rise of productivity in staple crops** • Plant genetic resources are considered as 'global commons'. • National seed collections linked to worldwide seed bank network to augment food production. • Private industry controls unauthorized propagation of plant varieties through technology (hybrids) and law (plant variety protection). • Worldwide replacement of landraces and wild plants by genetically uniform plant varieties.
THIRD 1980s-	**OECD countries strengthen position as major food exporters. Developing countries obtain niches in world market** • International food production chains dominated by agro-industrial conglomerates.	**Market-led industrialization of agriculture** • International treaties substitute the national state as main regulating agent. • Liberalization of agricultural trade.	**Rapid adaptation of plant qualities to diverse and changing world markets** • DNA sequencing becomes new conservation strategy for industrial conglomerates. • Private industry controls unauthorized propagation of plant varieties and the use of genetic information for the creation of new plants through technology (hybrids and 'terminator' gene) and law (plant variety protection and patents).

The conflicts that have occurred since the end of the 19th century on plant patenting, plant conservation, and in the direction of plant breeding research, suggest that crop development has been controversial throughout all Agro-Food Orders. Apparently, since the 1870s there have been differences of opinion about the way plant varieties should be improved in order to support national agriculture. The controversy gained an international dimension in the Second Agro-Food Order, when more countries began to participate in the international division of labour, and the social costs of regional and international differences in capacity in agricultural production and trade began to be felt worldwide. In the Third Agro-Food Order the controversy on crop development policies turned into 'the PGR conflict' as described in the introduction to this chapter. This conflict is not just national or international, but basically transnational in scope. There has been a considerable reduction in the national state's room for manoeuvre in designing a crop development policy in support of its national agricultural sector. Much of the regulating power has shifted to both intergovernmental organizations, such as the World Trade Organization, and to transnational corporations involved in crop development. Also, the political conflict over the rival agricultural production strategies has become transnationalized. Both advocates and opponents of a rapid industrialization of agriculture can be found everywhere, in OECD as well as in developing countries. This is the basis for the emerging transnational coalitions in the PGR conflict.

The book is organized in such a way that each chapter focuses on the conflicts over crop development in a specific historical Agro-Food Order. In Chapters 2, 3 and 4 we show that during the First and Second Agro-Food Orders the governments of several capitalist core countries provoked opposition when they facilitated the entrance of industrial capital in crop development. As agricultural production was largely organized on a national scale, conflicts on genetic information remained limited to national interest groups, e.g. between farmers, plant breeders and specialized seed producers. Chapter 5 illustrates how during the Third Agro-Food Order, the rapid expansion of private capital in agriculture has induced a transnationalization of crop development policy - and hence of the associated political conflicts. Because crop development increasingly depends on a ready access to specific genetic information, the rivalry between advocates of the alternative agricultural production strategies has provoked an intense conflict on the exploitation of genetic information. The final chapter examines the way export producers, domestic plant breeding organizations, commercial farmers and the peasantry in Chile and Colombia are embedded in this transnational conflict.

Conclusions

In a growing body of literature that documents or explains the controversy over the conservation and exploitation of plant genetic resources, the prime focus is on the international negotiations in which Northern and Southern states try to regulate the benefit sharing derived from genetic information. Also the few theoretical, social science accounts on the subject tend to separate the PGR conflict from its political-economic context and reduce it to a legal argument over international treaties.

We propose to assess the PGR conflict essentially as only one of the 'sparks' caused by frictions between what we call "rival agricultural production strategies". These strategies are based on different conceptions of how agricultural production should be organized. Since every strategy includes a particular vision about the role of new plant varieties in agricultural production, the differences also apply to crop development policies. The root of controversy on the exploitation of plant genetic information must be sought here.

Frictions between the production strategies not only differ per country but also vary throughout history. In order to analyse their historical dimension, we use the concept of the 'historical Agro-Food Order'. The purpose of distinguishing historical Orders is to illustrate that these frictions between agricultural production strategies are induced by two factors: (a) the degree of capital involvement in agriculture and crop development in a particular period, and (b) the political dominance of some of the actors involved in agriculture in that period. Hence, each Agro-Food Order produces its 'own' conflicts over crop development policies because of the diverging interests in the organization of agriculture.

Notes

[1] The term 'genebank' is confusing. The term has become the common, but incorrect, term for national and CGIAR-storage facilities that specialize in the conservation of populations representive of landraces, advanced cultivars, wild relatives of domesticated plants, and wild (i.e. non-domesticated) species, which are usually maintained in the form of plants, seeds, and tissue cultures. By far most collections consist of seed. As shown in Chapter 5, the term 'genebank' is more useful to describe recently established collections of cloned DNA sequences, some of which are indeed identified as 'genes'. In this publication we will therefore speak of a 'seedbank' when we refer to a storage facility of seed, plants, and plant tissue.

[2] The term 'plant genetic resources' was first commonly applied by the FAO Panel of Experts on Plant Genetic Resources during their series of technical meetings on the design of global conservation strategies in the late 1960s and early 1970s. The term referred to landraces, advanced cultivars, wild relatives of domesticated plants, and wild (i.e. non-domesticated) species used by man (Frankel and Soulé, 1981:181) but simultaneously emphasized the scientific importance and agricultural economic value of landraces and wild relatives. By consistently using the term in its reports, the Panel sought support within the FAO for the creation of the global conservation system. The lobby campaign of the Panel for a global plant genetic resources conservation system backfired on the agricultural scientific community when, in the early 1980s, critics adopted the term genetic resources to emphasize the North's economic benefits derived from the historical import of landraces and wild relatives from the South. In the present book, the term 'plant genetic resources' will be employed for genetic information in plants to which economic and political value is attached.

[3] In fact, the elimination of the 'common heritage' status of plants in the FAO preceded the CBD adoption. In anticipation of the conclusion of the CBD in June 1992, the signatory states to the FAO Undertaking in November 1991 agreed to include a third annex to the Undertaking in which the national sovereignty of nations over their plant genetic resources was recognized (Anonymous, 1991).

[4] In 1968, the American biologist Garreth Hardin published his often cited article "The tragedy of the commons" in which he contended that common natural resources are bound to become subject to over-exploitation if access and use were not managed (Hardin, 1968).

[5] The CBD was facilitated by an unprecedented number of nearly 1500 NGOs, which were officially accredited to the UNCED process. The NGOs encompassed civil, non-profit groups as well as industrial lobby organizations. Many of them were permitted to assist in the preparation of the national report of their governments, submit written statements, and make oral interventions during the official meetings of the preparatory and UNCED sessions (Doherty, 1994).

[6] In biology and plant breeding, all generations arising from a cross between genetically unlike parents are termed "hybrids". In the seed trade and in this publication, however, the term hybrid refers to the first generation of a cross between inbred lines.

[7] In this publication we only refer to agricultural production in market economies, because the processes of 'appropriation' and 'substitution' cannot take place in socialist economies. Even though in these economies agriculture has been greatly industrialized and many of its effects are similar to those in market economy countries, the dynamics of this industrialization is different. Since in socialist economies, the compelling need for capital accumulation is absent and the family farm has been replaced by alternative production units, the main actors and strategies for organizing agriculture have been different. The industrialization of agriculture in socialist economies consequently requires an alternative explanation, which is beyond the scope of this publication.

[8] Friedmann and McMichael (1989) also use the term "international food order" in their article, but without any explanation. The "food order" seems to be interchangeable with "food regime".

Chapter 2

The First Agro-Food Order (1870s-1930s) Crop Development to Improve Agricultural Competitiveness

This chapter discusses plant conservation policies and the legal protection of plants in the period between 1870 and 1930, to which we refer as the First Agro-Food Order. This period is characterized by a specific international division of labour in agriculture, consisting of the traditional colonial trade, complemented by a cereal trade flow from the settler states and Russia to Europe. The international division of labour gave rise to new forms of social regulation in agriculture. The competitive prices for which the American cereals were offered in Europe revealed the dissimilarity of production capacity among different countries. Farmers, landed aristocracy, and industry in most countries needed intervention by the government, particularly during periods of overproduction. In most core capitalist countries, Ministries of Agriculture were founded to organize agricultural production and crop development, with the objective of improving national competitiveness.

The ministries began to encourage plant breeding efforts and the conservation of seed in order to speed up the improvement of temperate crop varieties. The intensity of the support differed widely per country. The U.S. government, for example, undertook systematic efforts to collect foreign plant material, and was directly involved in plant breeding, while in various European countries, such as Great Britain and The Netherlands, the government was less dynamic. These countries still trusted the competitiveness of their colonial empire, and until well into the 20th century most public research continued to focus on tropical agriculture.

The changes in the organization of agriculture also raised legal questions. Private investment in crop development gave rise to private seed producers and seed distributors. The private plant-breeding firm also became a new economic and political actor that had to be reckoned with. Plant breeders suffered from the unauthorized exploitation of the new varieties they released. By comparing themselves with inventors, the new professionals demanded that the state grant patent protection on their plant varieties.

The conservation and plant patent issues are dealt with in the sections 2.2 and

2.3. We begin this chapter, however, with a discussion of the regulation of the seed market, one of the first progressive interventions of governments in Europe and the USA to facilitate the industrialization of crop development.

2.1 Agricultural competitiveness as a new political issue

The First Agro-Food Order is characterized by a major change in the international division of labour in agriculture. Towards the end of the 19th century, the newly independent settler colonies, particularly the USA and Russia, began to export bulk quantities of cereals to the European metropolises (Figure 2.1). These cereal exports initiated a pattern of agricultural trade that was fundamentally different from colonial trade.

The 'vertical trade' within the colonial empires was managed in the European metropolises and based on a complementarity of products that resulted from differences in climate and social organization. During the late 19th century, the metropolises imported tropical products from colonial plantations. Rubber, sisal, jute, indigo, and cotton were the raw materials in an emerging agricultural processing industry, while oils, tea, coffee and sugar were imported as luxury products for mass consumption by the growing industrial labour force. Production in the colonies, however, could not yield sufficient grain to feed the urban masses in the metropolises, and neither could the surrounding rural population in the metropolises, at least not at the price of the settler states. The settler states, with the USA in the lead, would become Western Europe's main grain suppliers (Friedmann and McMichael, 1989:97). This transatlantic trade flow implied that a new inter*national* market of temperate agro-food products was complementing the traditional colonial trade.

The emergence of a world market in temperate agricultural products through the cheap cereal exports to Europe revealed the competitiveness of American agriculture. The strong U.S. position was primarily due to the specific way in which American agriculture was industrializing. Land was readily available in the USA, but labour was not. Replacement of human and animal labour by mechanical devices was therefore the focal point of capital investment. This mechanization altered the organization of agriculture by reducing absolute labour requirements (especially for harvesting) and generated impressive gains in on-farm labour productivity. In Europe, mechanical equipment was adopted at a far lower pace due to the abundant labour supply, the small size of the farms (preventing the successful application of machinery) and the lack of credit facilities in rural areas. The development of commercial and capital-intensive agriculture in Europe would not depend on mechanical equipment, but on techniques allowing for a more efficient use of existing land (Goodman *et al.*, 1987:24).

Figure 2.1 Wheat exports by main producing countries (1851-1899) (in millions of bushels)

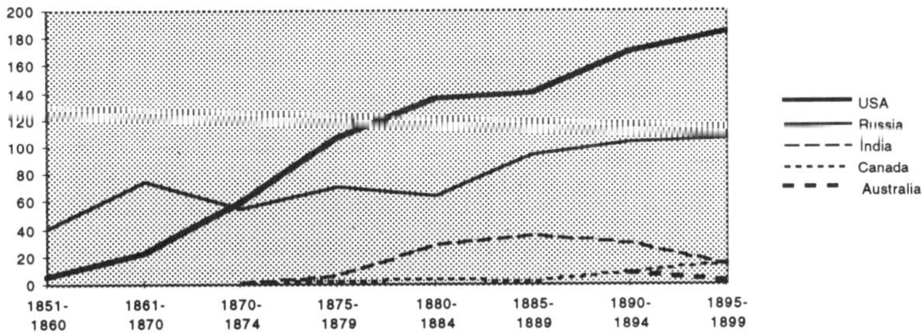

(Data derived from Tracy, 1982:21)

The import of cheap cereals induced a deep agricultural crisis around 1870, which affected all European metropolises, prompting them to intervene in the agricultural sector. In those countries where landed aristocracy and farmers could generate enough political strength vis-à-vis industrial interest groups, such as in France, Germany and Italy, the government erected tariff walls for the protection of its national producers (Gourevitch, 1986). More progressive measures to improve the national competitiveness in food production were also considered, and led to the design of crop development policies for temperate agriculture in most European countries.

The first step taken in Europe involved the regulation of seed supply. An increase in the demand of seed in the 19th century had created a powerful stimulus for farmers, traders and governmental institutions to become involved in plant selection, introduction, and conservation, as well as the production and distribution of seed. The result was a wide diversity of plant varieties in the farmers' fields: old landraces, newer regional farmer selections, new selections that were released by specialized private enterprises, and varieties bred by scientists at public research institutes. Foreign seed enlarged the plant diversity as soon as improvements in transport and communication systems enabled commercial seed to be traded inter-regionally and internationally.

This situation, which was referred to in Europe as 'varietal jumble' facilitated fraud and swindle. Inferior seed was misleadingly offered under the name of a reputable variety and high quality seed could be mixed with worthless weed seed. Fraud in seed sales could be rewarding because the quality of the seed was assessed only months after the purchase, by which time it was too late. Farmers used to rely on the reputation of "the man behind the seed" (NTZ, 1992:42), but the growing number of varieties and seed sellers at the end of the 19th century made this rule obsolete.

Reliable seed is the basis for a competitive agricultural sector. The chaotic situation in seed supply at the end of the 19th century hampered the rise in agricultural production that was required to feed the expanding urban industrial labour force, and to meet foreign competition. The uncontrolled seed market did not only upset farmers and landed aristocracy; it was also resented by breeders and seed production firms who were trying to establish a business. This was the reason that in the major European countries a political majority accepted governmental control of the seed market. The basis for seed regulation was laid in the 1870s, when governmental seed testing stations emerged in most West European countries (Addens, 1952:52). In the early 20th century, they extended their involvement in agriculture to seed certification, which gave farmers a guarantee about some essential qualities of the seed they purchased.

Far more than European governments, the U.S. government had traditionally played a pivotal role in plant introduction and seed distribution. Like other settler states, the USA was poor in native genetic resources. The seeds that European settlers had brought with them performed poorly in the new country.

> "[T]he wheat and small grains that were intended to grow in America failed so miserably that, except for their discovery of Indian corn, everyone would have starved." (Crabb, 1947:7).

Thus the introduction of new plants from all over the world and seed supply within the USA had been a political priority for a long time. In order to speed up the distribution of seed, the Agricultural Division of the U.S. Patent Office set up a 'free seed' programme as early as 1839. Through this programme American farmers were stimulated to grow a large variety of plant material and to report back the results. In 1897, over 1.1 billion seed packages had been freely distributed by the state (Busch *et al.*, 1995:76; Kloppenburg, 1988a:60-63). The free seed distribution, however, was direct competition for the commercial seed trade of the private seed firms that had entered the market. The private industry was increasingly concerned with the state's involvement in seed supply. In 1883, the seed industry founded the American Seed Trade Association (ASTA) and set itself the task of eliminating free seed distribution and amending seed legislation so that it would facilitate private capital investment in crop development. The free seed programme was terminated in 1924 (Kloppenburg, 1988a:73).

2.2 Using Mendel in improving agricultural competitiveness

The emergence of an international market for temperate agricultural products was closely related to the rapidly increasing demand for food for the labour forces in urban areas. In particular, grains and other staple-crops (such as potatoes) were required because they formed the basis of the traditional diet, were relatively high in protein, and were easily storable and transportable. European countries could meet the growing urban demand with grain imports, but these also made them increasingly dependent on world market prices. The USA on the other hand had become too dependent on wheat exports, as it first experienced during the wheat crisis in the 1880s.

It was against this background that crop development became a national priority, particularly in the USA. State involvement in crop development induced biological research and enabled the rediscovery of the findings of Gregor Mendel which eventually formed the basis for a new paradigm in biology: 'Mendelian genetics'. The new insights into heredity would lead to a change in research criteria in crop development research that eventually proved to be of great importance for the international competitiveness of the USA. During the heydays of the colonial powers, crops were improved by selecting and propagating individual plants that had a better overall performance. This strategy required the continuous collection of individual plants in various places of the world, a strategy which is known as 'botanical exchange'. The benefit of Mendelian genetics was that it enabled the targeted transfer of specific characteristics from the one plant to another through crossing. With the change in research strategy, the focus of plant collection shifted accordingly, from whole plants to plant characteristics.

The speed and intensity with which these changes occurred differed per country, however. There was a notable difference between the USA, on the one hand, and the European colonial powers, on the other. The U.S. plant introductions immediately served American competitiveness in temperate agriculture, whereas the plant improvement activities in the European metropolises remained predominantly focused on the competitiveness of tropical agriculture in their colonial territories.

2.2.1 The end of the botanical chess game

For centuries, plant collection and introduction of new plants predominantly took place within and among the colonial territories. A key role in this collection and shipping of plants was played by botanical gardens, of which the Royal Kew Gardens in London is perhaps the best known. Kew Gardens was established in 1841 and formed the centre of a virtually global network. In 1889, the number of 'satellite gardens' attached to Kew was no less than 56, spread over South and Central America, West Africa and East Asia as well as the settler states. The gardens contained hundreds of thousands of (potentially) useful plant varieties of new tropical plantation crops.

Plants were transferred within botanical networks for two reasons. In the first place, throughout the 19th century, plant improvement in the colonies mainly involved the selection of individual plants on the basis of visible characteristics (Sneep *et al.*, 1979:191,204). The resulting 'new' varieties were propagated in infinite numbers. The combination of selection and propagation was agronomically and economically precarious, however. Plantations were filled with the direct descendants of very few crop plants which, because of the limited genetic variation, were highly susceptible to pests and diseases.[1] Hence, a constant exchange of plant material within colonial botanical networks was necessary to reduce susceptibility to local pests and diseases. If the material required could not be found within the network, adventurous voyages to rival colonies or non-colonial territories were required. In addition, the risk of infection by pests and diseases required the botanical network to organize transfers of plant material to remote geographical areas where they were not endemic and where natural enemies were absent.

The other main reason for plant transfers was related to the immobility of labour force. In spite of the slave trade, labour was not always located in the countries where plants with economic potential originated. Only in a few exceptional cases, both the enslaved or indentured labour force and the plants could be shipped to new areas, e.g. in the case of sugar production in Cuba. Promising plant material was thus constantly and systematically shipped from one garden to another, a practice Lucile Brockway has coined the "botanical chess game". Brockway (1979; 1988) has extensively described how Kew officials searched for or stole useful plants, particularly in Latin America, and how they selected the best plants and transferred them to Asia. The Kew botanical garden thus played a central role in extending the plantation system to Asia. By the mid-19th century, Kew had perfected its taxonomic knowledge with the help of botanists, such as Charles Darwin. Other colonial powers had institutes equivalent to Kew Gardens. France had the *Muséum Nationale d'Histoire Naturelle*[2], Germany the *Botanische Zentralstelle für die Deutschen Koloniën*[3], the Netherlands the *Botanische Tuin van Leiden*, Spain the *Jardín Botánico del Soto de Migas Calientas*, and Sweden the *Linneaus* garden.

As access to good plant material was essential for plantation economies in the colonies, rival colonizers, predominantly the Dutch, Spanish, Portuguese, and French, were eager to get hold of each other's material. The quest for good plant material could have a profound impact on the inter-colonial competition, as is shown in the often-quoted case of the 'rubber coup'. Access to rubber became of vital importance in the late 19th century, because of the emergence of motorized vehicles in Western Europe and the USA. After several unsuccessful attempts, Kew officials in 1876, managed to smuggle 70,000 rubber seeds out of Brazil, at that time the world's largest rubber exporter. The seed was transported via the Kew Gardens network to Malaysia, where they became the basis of a newly established rubber industry that challenged the traditional producer. By 1919, the Brazilian rubber industry was virtually dead (Brockway, 1979:141-165; De Souza Silva, 1989:213-222).

The example of rubber illustrates the profound impact that individual plants and related botanical knowledge had on the colonial division of labour. The success of plant transfers could make or break plantation economies. Botanical exchange and military power were closely related. The export of plant material was often forbidden and controlled by military means. Botanical knowledge could also support military interests. In 1820, it was discovered that alkaloids isolated from the bark of the *Cinchona* tree, called 'quinine', could help to cure and prevent malaria. Malaria was one of the most devastating diseases among troops stationed in tropical regions, particularly West Africa and India. The growing of *Cinchona* was able to start in India after Kew officers had smuggled seeds out of the Andes region (mainly Peru and Bolivia). A British monopoly was not to emerge, however. The Dutch on Java managed to obtain *Cinchona* seeds from an English trader. On the basis of these seeds, the Dutch managed to establish a monopoly on quinine[4] by

1890, forcing many planters in India and Ceylon to turn to coffee and tea (Brockway, 1988:55-57).

2.2.2 The discovery of the gene

Ever since men started to plant and harvest crops, plant characteristics have been adapted to serve specific aims. By continued selection of promising individual plants generation after generation, better varieties were developed. By experience it was known that selecting for visible characteristics (such as size or colour) was 'easier' than invisible features (such as resistance to pests and diseases). Similarly, plant developers knew that improving self-pollinating plants had a more predictable outcome than cross-pollinated plants, because the latter were liable to be polluted by unwanted characteristics transferred from other plants. In the 19th century, experts in 'economic botany' were the main actors in crop development in the colonies. Trained in the metropolises, they spent most of their time in the colonies selecting and 'discovering' new species and varieties.

Although at the end of the 19th century, selection of plants had become less an art and more a profession, plants remained 'black boxes' as far as inheritance was concerned. Biologists in Western Europe, equipped with increasingly advanced microscopes, had been able to understand how cells and their nuclei divided and reproduced. As early as 1883 it was understood that the mechanism of inheritance was to be found in a 'hereditary substance' called "germ plasm" located in the nucleus of a cell (*cf.* Stubbe, 1971; Mayr, 1982:679). At about the same time it was understood that minute threads, called "chromosomes", contained this hereditary substance. But the big question, especially for agriculture, remained how the process of inheritance precisely worked. Breeders knew that the answer to this question would allow for a targeted transgression of both visible *and* invisible characteristics (such as disease resistance) from one plant to another.

In the context of his evolutionary theory, Charles Darwin had been puzzled by the mystery of inheritance almost all his life. In 1868 he concluded that two types existed: "soft-" and "hard inheritance". Soft inheritance supported the idea that plants could, and eventually would "acquire external characters". In this idea, climate, nutrition, or other environmental factors could change the genetic constitution of a plant. The theory of hard inheritance, suggested the opposite: the genetic make-up of a plant would not, or would hardly, be affected by its environment. The theory was based on the premise that plant characteristics constituted a "fixed set of factors", but that environmental factors could change the *expression* of these characteristics (Mayr, 1982:687).

With the publication of Factors of Inheritance in 1866, Mendel was the first to discover the mechanism of hard inheritance. It was, nevertheless, only in 1900 that his theory was fully appreciated. In that year, Hugo de Vries (The Netherlands), Erich von Tschermak (Austria), and Carl Correns (Germany) carried out experiments very similar to Mendel's, and while their "work was well under way" they simultaneously became acquainted with Mendel's paper of 1866 (Stubbe, 1971:265). The re-discovery of Mendel's laws caused a revolution in biology. Between 1900 and 1910 an international scientific 'gold rush' took place in which scientists in virtually all West European countries, the USA and Russia searched for a theory which would connect the phenomenon of chromosomal division to Mendel's laws of inheritance.

The British scientist R.H. Biffen, in his paper "Mendel's Laws of Inheritance in Wheat Breeding" of 1906, was the first to apply Mendel's work to crop development (Robinson, 1996:16). Biffin's discovery was truly seminal as it not only showed that resistance to a described disease was inherited in a Mendelian fashion, but also that this process was controlled by individual "factors of inheritance" carried on a chromosome. In 1909 these factors were coined 'genes'. This discovery gave plant breeders a very precise selection criterion for plant improvement. The discovery of the gene is generally regarded as the starting point of modern genetics and even of modern scientific breeding. Subsequent progress in genetics was made at a different pace in different countries. Great Britain, without any question, assumed leadership in Mendelian genetics in the 1910s and 1920s, soon to be followed, and eventually overtaken, by the USA (Mayr, 1982:731-734).

Mendelian genetics offered new opportunities for plant development. Rather than laborious selection work on the basis of trial and error, the development of new plant varieties became a scientific activity with a more predictable outcome. 'Scientific breeding' can be considered as a breeding method in which individual plant characteristics can be identified and selectively transferred by crossing two plants. The basis of this practice was laid in 1909 by a Danish scientist, W.L. Johannsen, who developed 'pure lines'[5] of self-pollinating crops. The economic significance of pure line breeding was tremendous. While Mendelian breeding allowed for a controlled mixing of genetic characteristics, pure line breeding offered a practical method to 'fix' them in succeeding generations.

The opportunities for more effective plant breeding, offered by Mendel's laws, required a plant and seed introduction system that differed from the botanical exchange system. Plant material that would have been of value for selection in the colonial period could, according to the new genetic criteria, be ignored. Rather than collecting individual plants that were selected for their overall performance, the collector's task was to supply breeders with a useful 'genepool', a group of genetically related plants that carried various characteristics, some of which may be useful for and transferable by breeders. The application of Mendelian genetics on plant improvement induced "gene hunting" (Juma, 1989) and implied a tremendous expansion of plant collection activities.

The search for genes relied on an even more sophisticated apparatus than the botanical network. Exchange of plants and seeds containing potentially useful germplasm among breeders became much easier, because professional breeders emerged with a comparable scientific background in Mendelian genetics, who started to communicate outside the traditional colonial institutional settings.

Furthermore, in contrast to the importance of individual plants in the botanical system, the exchange of seeds and plant material containing specific known genetic characteristics was less of a threat for agricultural production. Where single plants could make or break an entire tropical plantation economy, genes rather became one of the inputs of crop development. Hence, seed and plant material with agronomically useful 'germplasm', the physical expression of chromosomes and genes, started to circulate among international scientific communities that were less constrained by political barriers.

Below we explore how the European metropolises, their colonies and settler states attempted to benefit from modern genetics by adjusting and implementing research and conservation policies. Although they were implemented practically everywhere, the impact on food production as a whole differed substantially. As we will show, the USA was far more successful in integrating modern genetics into their research and conservation policies than the European metropolises.

2.2.3 Public research system increases U.S. competitiveness in agriculture

After the 1870s, large-scale and mechanized production, and the widespread distribution of improved plant varieties, allowed the USA to become the world's largest wheat producer and exporter (Figure 2.1). The role of the U.S. government had been instrumental in achieving this leading position. As one of the first countries, the USA, in 1863, established a Department of Agriculture (the USDA). A little later, the State Agricultural Experiment Stations (SAES) were founded. They were complemented by the so-called Land-Grant Universities (LGUs), which taught modern genetics to agricultural scientists (Busch *et al.*, 1995:79).[6] In 1878 the USDA was spending a third of its budget on the collection and distribution of plant material (Kloppenburg, 1988a:60).

Considering these vast investments in agricultural research, the question arises why it took four decades before Mendel's publications were appreciated by agronomists. Mayr's (1982:722-726) explanation is that Mendel was so far ahead of his time, that other biologists were simply not able to understand the relevance of his findings. We propose an additional explanation that attributes the 're-discovery' of Mendel to the changing international division of labour in agricultural production. Mendel's laws were re-discovered in Europe at a time when colonial tropical agriculture was still unrivalled. This allowed European scientific circles, particularly at Cambridge University in Great Britain, to treat Mendel's laws as 'pure science' and difficult to apply. In the USA, however, there was an explicit need to employ the new insights to crop development. American farmers were just recovering from the wheat crisis. It was therefore the USDA, and not a European agricultural research centre, which first attempted to apply Mendelian principles in order to strengthen national agricultural production. Soon "news of Mendel's work and the experiments and controversies it prompted was available to experiment-stations in every state." (Paul and Kimmelman, 1988).

The re-discovery and application of Mendel's laws put U.S. agricultural research and seed collection in a higher gear. The main impetus came from the government around 1900 when a new special USDA unit, the Office of Foreign Seed and Plant Introduction, was created. In the following twenty years, four Federal Plant Introduction Stations were built, which would store the hundreds of thousands of plants and seeds that were collected in the subsequent century (Hyland, 1961:471,478-479). The stations still perform a key role in U.S. collection and conservation.

Sustained support for public agricultural research came not only from the government. The country's most important industrial capital groups, such as the railroad companies and banks, also had an interest in expanding public agricultural research.

> "[A] deteriorating agricultural trade balance raised fears as to the capacity of the United States to sustain its position in international trade. The Country Life Movement is probably the most widely known and studied response to what was perceived as a problem of national significance. Organized around prominent agricultural scientists such as Cornell's Liberty Hyde Bailey and Gifford Pinchot of the USDA, the movement was funded and supported by urban-based interests and industries that depended on the flow of agricultural commodities for their revenues. What these groups shared was an interest in rationalization of agriculture through science. If the agricultural scientists were motivated by a vision of a transformed and improved rural society, the business interests were unambiguously interested in restoring productivity advance as a necessary condition for their own continued capital accumulation." (Kloppenburg, 1988a:74)

Despite the institutionalization of a federal agricultural research and extension system, it took some decades before the knowledge of Mendelian genetics could generate large improvements in disease resistance. Research remained focused on selecting plants for yield and adaptation (Hyland, 1961:472). It was only in the 1930s, during the advent of the Second Agro-Food Order, that experiments in scientific breeding began to be translated into significant increases in productivity, especially in maize.

2.2.4 Changing interests in Europe: from tropical to temperate agriculture

The European metropolises had great difficulties in competing with the cheap American grain. They constituted the political and economic heart of colonial powers and were preoccupied with the competitiveness of their tropical agriculture in view of inter-colonial rivalry. Inter*national* competition in temperate agricultural production in the home countries was a new phenomenon for which a response was not immediately found. Research efforts in temperate agriculture remained limited in comparison to that in the tropical territories. Numerous experiment stations were established after 1850 to sustain the production in the colonies of export crops such as sugar, tobacco, coffee, tea, spices and fruits. Of the French empire's 115 experiment stations, only 40 were located in France itself in 1930. At that time, the British empire had 450 experiment stations, of which only 30 were located in the home country (Busch and Sachs, 1981:134-135). These stations replaced the botanical gardens as central actors in tropical plant improvement. Production of staple crops in the colonies to supply the urban labour in the metropolises hardly occurred, as no colony was able to compete with the cheap grain production of the settler states. The British considered an increase in wheat production in India only after the opening of the Suez canal in 1869, which allowed Indian wheat to be marketed in Europe at competitive prices (Figure 2.1) (Busch and Sachs, 1981:136).

While most of the tropical research stations were state-funded, the sugar stations in Java, part of the Dutch East Indies, were in private hands. The crisis in the sugar industry and the subsequent *sereh* disease in 1883 instigated the Dutch sugar planters to invest in research in order to protect their market share. They built three experiment stations which would merge into the Java Sugar Industry Experimental Station (KAW, 1923:13-14). In the mid-1920s, this private sugar station was the largest of its kind in the world, with an annual budget of US$ 500,000 (Busch and Sachs, 1981:138).

Even though the new, tropical experiment stations were concentrating on a single crop and were staffed by botanists, Mendelian genetics was hardly applied. The stations continued to rely on 'economic botany', i.e. crop development through selections of introduced plants on the basis of taxonomic and botanical knowledge (Busch and Lacy, 1981; KAW, 1923; Masefield, 1972). Real applications of pure line breeding were not available in the tropics until the late 1920s. As a consequence, the need for a change of plant collecting strategies remained absent.

In none of the West European countries was the new knowledge employed to increase competitiveness in temperate agriculture. This is surprising since modern genetics originated in Europe and was also first applied to wheat, a temperate crop. The reason for the incompatibility of developments in Mendelian genetics and crop development, at least in comparison to the USA, may be sought in the attitude of the European governments, which remained relatively aloof in respect of agricultural research.

In *France*, "not much happened until the 1930s" in terms of studies on genetics (Mayr, 1982:732). In the late 19th century the annual agricultural research budget in the USA was already 20 times that of France (Paul, 1985), despite the fact that promising developments in French plant improvement research had occurred in the mid-19th century. For example, the Vilmorin brothers had developed at that time a revolutionary breeding method which replaced the traditional more imprecise mass selection technique with that of "family selection" (Sneep *et al.*, 1979:204).

In *Great Britain*, a Parliamentary decision in 1909 to revitalize and modernize agriculture had resulted in the foundation of four primal agricultural research institutions[7] (Palladino, 1996:118-119). Two of these institutions were later headed by W. Bateson and R.H. Biffen, leaders in the new field of genetics. But in spite of the early theoretical insights and the research capacity available, Great Britain could in no way measure up to agricultural research in the USA. Since there was little agreement among governmental, private and academic circles about the importance of the Mendelian breeding for plant breeding, no real plan existed that linked agricultural research to agricultural production. The research stations operated more or less on their own, both financially and intellectually. In the period 1920-1929, only 40% of the funds for plant breeding research came from public sources (Palladino, 1996:124). As far as practical usage of Mendelian insights was concerned, breeding work focused on quality improvement. In 1916, Biffen managed to produce a 'hard wheat' variety that could compete with wheat varieties imported from North America. Biffen's wheat, however, could not curb the transatlantic wheat flow. At the University of Reading, where agricultural science was traditionally put more at the direct service of agricultural production, Mendelian genetics was dismissed as a useful basis for plant breeding (Palladino, 1993:312,319). The first British state-funded breeding research project would occur only after the Second World War (Thirtle *et al.*, 1997:24,37).

In *The Netherlands,* most crop development activities have been carried out by private seed producing family firms and farmer-breeders. Renowned are the achievements of farmer G. Veenhuizen, who developed numerous new potato varieties[8] (Addens, 1952:93,100). Public money for agricultural research became available in 1912 when the national Plant Breeding Institute (IVL)[9] was established in Wageningen. The IVL was added to an existing network of provincial experiment stations. In contrast to most other Western European countries, the Dutch public research centres merely enforced existing private breeding activities rather than forming an alternative to them. The IVL predominantly monitored existing cultivars, most of which had been produced by private breeders (Addens, 1952:93-95). The Dutch government's limited involvement in agricultural research was directly related to a traditional non-interventionist policy in agriculture in The Netherlands itself. The IVL was financed by the Directorate of Agriculture, which was part of the Ministry of Water Management, Manufacturing and Trade. A genuine Ministry of Agriculture was founded only in 1935.

Germany held a somewhat exceptional position in Europe. The country was the forerunner in the agricultural experiment station movement during the second half of the 19th century. In 1900, 75 stations were operational in Germany and in its African colonies (Busch and Sachs, 1981:134). Governmental support for crop development was expanded after the First World War. The Germans had suffered seriously from the food blockades during the War and urgently required quality seed. In order to become more self-supporting in food, the Research Institute for Plant Improvement[10] was founded in 1927, while a handful of regional seed research facilities were built to improve seed supply. In the 1920s, at least five important botanical expeditions were made to Turkey, Spain and South America for the collection of species of wheat, maize, and various other temperate crops (Flitner, 1995:41,49,278). In spite of these investments, German research in the early 20th century continued to work in the tradition of the 1880s (Mayr, 1982:732) with a strong emphasis on (advanced) selection methods. In 1914, Erwin Baur, Germany's most prominent geneticist, would still conclude that "it will take years of intensive labour before practical plant breeders can also start working with cross breeding on a large scale" (translation RP/JvW) (Flitner, 1995:41). Public spending on agricultural research and modern genetics would expand only in the 1930s (see section 3.3).

Thus, governmental interest in agricultural research in Europe was relatively weak compared to the USA. Mendelian genetics was hardly related to crop development, and consequently plant-collecting activities hardly changed until the 1930s. The main task the governments assumed was that of testing foreign plants and seeds. Throughout the 19th century, the introduction of new plants in Europe, however, remained an arbitrary activity of often large farmers with foreign connections.[11] The exception was Germany, whose government was heading for food autarky. German collection activities were resumed on a large scale shortly after the First World War had ended.

Summary

During the First Agro-Food Order, a shift took place from intra-colonial trade based on the complementarity of products resulting from different climates, to inter*national* trade among core capitalist countries in temperate zones. Concern for national competitiveness caused states to become involved in crop development. This formed a major prerequisite for the re-discovery in 1900 of several insights into the heredity mechanism of living organisms. The 'promise of Mendel' was that breeding based on genetic principles could enhance the production potential of temperate crops in the core industrial countries.

Between 1900 and the 1930s, Mendel's laws were gradually worked out within academic circles on both sides of the ocean. The USA would be the first to employ genetics for the rationalization of its crop development policies and infrastructure. Incentives came both from the U.S. government and from industrial organizations. While the government's direct intention was to stimulate diversification in order to curb overproduction in wheat, the industry saw opportunities in increasing food processing and transport. In Europe, crop development did not receive such firm public or private backup. The European metropolises were still more directed towards the inter-colonial competition of tropical agriculture. During the First Agro-Food Order, they shifted their research focus to temperate agriculture, but did not keep the pace that was set in the USA.

The new breeding paradigm had a profound effect on conservation strategies both in Europe and the USA, but again the impact was different. Although the metropolises continued to rely on the intra-colonial exchange of tropical botanical plant species, botanical exchange became less relevant for the new breeding paradigm. This relevance further decreased because the application of Mendelian breeding techniques on tropical crops, many of which are non-seed propagated, is much more complicated than with seed propagated plants. The metropolises long experience with botanical exchange could not guarantee a successful switch from botanical exchange to plant and seed introduction based on scientific criteria. Not so in the USA, where initiatives were taken to tie the new research strategies to conservation of temperate seed and plant material.

2.3 Are plant breeders inventors?

The industrial revolution had shown how important technological progress was for economic development. With the expansion of the world market it was also realized that the continuous generation of new techniques was essential for achieving international competitiveness. Thus, support of inventors became a priority for governments both in European and settler states in the 19th century. The major policy to nurture technological innovation was the introduction of patent law.

Today it is often thought that the protection of plant varieties was among the objectives of that patent system. As soon as farmers became breeders specializing in the development of new plant varieties, they had to cope with competition from farmers, seed producers and other breeders, who freely propagated a new variety and used or sold the seed. The commercial problems arising from this competition were comparable to those of inventors, composers and authors. The state did assist them in curbing the unauthorized reproduction of their work by granting patents and copyrights. The emerging crop development industry envisaged similar protection for their plant creations.

In this section it is shown that plant patenting occurred only at the very end of the First Agro-Food Order. There was protection for fruits and flowers, crops in which a strong private sector had emerged, but which had little to do with innovation and national agricultural competitiveness. Incomplete botanical knowledge and insufficient technical control over biological reproduction processes of plants not only hampered private investment in crop development of staple crops, but also plant patenting.

Below we deal with the legal protection of plant varieties during the First Agro-Food Order from three angles. First, we examine the waning mercantilist form of plant protection that was employed by the European colonial powers. Then we examine the political dimension of patent law, and the initial plant patent initiatives by private American and European plant breeders in temperate agriculture.

2.3.1 Colonial botanical monopolies

For a long time the colonial economies had been highly concerned with the protection of individual plants of crops on which the plantation agriculture was based. As indicated earlier in this chapter, all European colonial empires collected and stole botanical species even in the most remote parts of the world. But as far as their own plant stock was concerned, they sought the protection of the 'botanical monopoly'. This monopoly refers to a ban on the export of propagating plant material, enforced by the government. Even though widely used, the botanical monopoly proved to be an instrument of little effectiveness. Only in some cases was it successful during a short period. Spices and cacao plants had a relatively small and circumscribed natural habitat, and had not been widely diffused when the Europeans came upon them (Brockway, 1979:57). Nevertheless, the botanical monopoly remained in use until the late 1920s.

To illustrate the attitude of plant breeders and authorities as regards plant protection in the European colonial territories, we will briefly examine the Dutch sugar cane industry in Java. As mentioned earlier, the Java Sugar Industry Experimental Station was the largest private sugar experiment station in the world in the mid-1920s, with an annual budget of half a million dollars. The private breeders raised the question of unauthorized diffusion of newly released sugar cane varieties in the Java industry in the early 20th century.

The sugar breeders tried to raise their income by controlling the unauthorized use of their new varieties within the colony through purchase agreements. These agreements prohibited purchasers from distributing propagating material of the new variety among third parties. They also established a mutual and remarkably 'progressive' remuneration system in 1924.[12] Several registrations were filed under the system, but the quick expansion of the successful variety POJ 2878 was the reason that few other clones were used. Royalty payment to other breeders was therefore very modest (Van Harreveld, 1934:123,126-128).[13]

The mutual remunerating system only served a local function among the Dutch plantations in Java. For protection beyond its own circle, the industry relied on governmental support. In order to curb the diffusion of modern varieties among the indigenous population of Java, the Dutch sugar interests were protected through a police ordinance which prohibited local communities from using modern cane varieties developed by the Dutch. The indigenous population was only allowed to grow and market their landraces (Van Harreveld, 1934:112).

Foreign diffusion of propagating material of sugar cane and tobacco was restricted through a self-imposed export ban for which governmental backup was demanded. After the Java sugar clone POJ 2878 had attracted the attention of rival colonial empires because it raised annual sugar production in Java by 24%, the Dutch authorities started to think of methods to prevent the use of the new varieties in other colonies. Are the Dutch Indies "suffering from their fame?", wondered the local newspaper, the *Soerabayasch Handelsblad* in 1928. How could the new *Hevea*-clones be prevented from spreading throughout Malacca, and the Java sugar cane clones from spreading all over the world? (quoted in Van Harreveld, 1934:112).

The Dutch colonial authorities had already prohibited the export of tobacco seed in 1924. The government's response to the sugar interests came some years later. In 1927 the International Convention for the Abolition of Import and Export Prohibitions and Restrictions was ratified by the Dutch government only as far as Europe was concerned, thereby enabling the colonial authorities to ban the export of tropical plant material (Van Harreveld, 1934:129).

Protection of plants *as intellectual property* was apparently not relevant in inter-colonial competition. When in 1928 the sugar industry demanded an export ban on individual sugar canes in the People's Council, the authorities referred to intellectual property regulation, which was in the air.[14] But such protection was not desired by the sugar industry. A member of the Council stated:

> "Getting a high price for improved planting material is not what we want; we want to prevent its use by others, and that will never be possible under the measures for the protection of scientific property." (translation RP/JvW) (*Handelingen Volksraad* quoted in Van Harreveld, 1934:129).

The colonial sugar industry's outcry resulted from its concern that it was relatively easy for rival colonial industries to acquire the precious Dutch sugar varieties. Receiving royalties for the propagation of the sugar varieties was not relevant. The protection of the Dutch competitiveness in the sugar trade required protection against any exploitation of sugar cane varieties, be it for propagation or for breeding purposes, in other colonies. In the 1920s such protection could only be achieved by means of a total ban on the export of planting material.

2.3.2 The European patent controversy

The botanical monopoly as a mechanism for protecting tropical plant varieties and colonial competitiveness withered away during the First Agro-Food Order. Meanwhile, the changing international division of labour gave rise to new forms of protection, directed at temperate agriculture. The international patent system that was founded at the end of the 19th century harmonized national patent laws and raised the interest of plant breeders in the protection of their new creations by patents.

A patent[15] involves the right granted by the state to an inventor, which gives the latter the legal opportunity to prevent unauthorized commercial use of his invention for a specified number of years. In order to be eligible for a patent, the invention has to fulfil specific requirements; for example, that it has to be novel worldwide. The patent is of European origin. The oldest examples of grants of exclusive rights by kings and rulers to private inventors to practise their new arts or skills go back to the 14th century, to the Republic of Venice, but spread in subsequent centuries to other countries on the continent and to Great Britain.

One of the features of the patent is its national scope; it can be enforced only within the territory of the state that has granted it. This limitation is disadvantageous for countries that export technology. Having no patent protection abroad, an exporting firm runs the risk of having its innovations freely imitated by foreign competitors. From the perspective of countries that import technology, thc opportunity of imitation may be attractive, and may induce a state to deliberately refrain from enacting a patent law. Because of its impact on technology diffusion, patent regulation is not just a legal but also a political tool in the hands of governments.

In the mid-19th century, with the industrial revolution in full swing and the international market expanding, the role of technology in national competitiveness grew. It was in this period that for the first time patent law was considered as a national trade policy issue. Between 1850 and 1875, the patent system was intensely debated throughout Europe. It was a political controversy between free traders and protectionists, but it was also academic in character, as scholars tried to find a philosophical justification for patent law (*cf.* Machlup and Penrose, 1950:28). The academic side of the patent dispute remained inconclusive; the political side not.

At issue after 1860 was the establishment of a convention that would provide a minimum standard of patent protection among the then industrializing countries. The convention was a priority for the advocates of a protectionist trade policy, but was opposed by an anti-patent movement of 'free traders'. The latter comprised a host of individuals and organizations across Europe, including trade associations, Chambers of Commerce, organizations of economists and the influential London *Economist*, who opposed patent protection as a new type of monopoly or tariff protection. Many liberals were concerned that patents would hamper the prompt general utilization of useful inventions, although some of the classical English economists, among them Adam Smith, Jeremy Bentham and John Stuart Mill, defended the patent system (Machlup and Penrose, 1950:3-5; Machlup, 1958:19).

The anti-patent movement was strongest in two countries with a relatively late industrial development: Switzerland and The Netherlands. The former lacked a patent law throughout the 19th century, the latter had its inadequate 1817 patent law repealed in 1869. The 'patentless' period had a beneficial impact on the development of some industrial sectors in both countries (Schiff, 1971).[16]

Only a few years after the abolition of the patent system in The Netherlands, the patent advocates all over Europe started a massive counteroffensive to prevent their countries from following the Dutch example:

> "New societies were formed, resolutions were drafted and distributed to the daily press, speakers were delegated to professional and trade association meetings, floods of pamphlets and leaflets were released, articles were planted in trade journals and reproduced in daily papers, public competitions were announced with prizes for the best papers in defence of the patent system, petitions were submitted to governments and legislatures, international meetings were arranged and compromises were made with groups inclined to endorse liberal patent reforms." (Machlup and Penrose, 1950:6).

The idea of patent protection in Europe regained its public appeal when, after the crisis of 1873, protectionists won out over the free-traders in many countries. Instrumental to this was the "strategic compromise" between patent advocates and opponents on compulsory licensing (Machlup, 1958:5). Inclusion of this principle in patent law offered the state the opportunity to break patent monopolies if patentees abused their rights and in case the public interest was at stake.

Patent law could now firmly take root in all major countries. The drastic patent reform bill that had been passed by the British House of Lords was withdrawn in the House of Commons in 1874. In Germany, a patent law for the entire *Reich* was adopted in 1877. Switzerland adopted a patent law in two phases, in 1887 and 1907, and also Japan, which had adopted its first patent law in 1872 only to abolish it again in 1873, enacted another law in 1885 (Machlup, 1958:5). Finally, The Netherlands, "the last bastion of 'free trade in inventions'", reintroduced a patent law in 1912 (Machlup and Penrose, 1950:6). The victory of patent advocates was sealed by the conclusion, in 1883, of an international treaty, the Paris Convention for the Protection of Industrial Property.[17]

2.3.3 Protection of plants under the Paris Convention

Today it is generally thought that the protection of plants by patents was already sanctioned in 1883, when the Paris Convention was adopted. This assumption would be confirmed by the Convention's explanatory note, which explicitly refers to agricultural products. In reality, however, the inclusion of agriculture was related to trademarks, not to patents.

The objective of the Paris Convention was to constitute minimal international unification in the protection of 'industrial property', covering patent protection of inventions, and protection of tradenames, trademarks, industrial designs, utility models, service marks, and indications of source or appellations of origin. These forms of industrial property suggest that the term industrial only refers to manufacturing activities in the factory. During the preparatory meeting in 1880, this provoked the Portuguese delegate to propose the explicit reference to "industrial activity in all classes", so that agricultural products were also eligible for protection. The delegate said that there was considerable counterfeiting of agricultural products, especially wines (Bureau International de L'Union, 1902:27-28). In response to this request, the closing protocol of the 1883 Act provided that:

> "The words 'industrial property' have to be understood in a broad sense and apply not solely to the products of industry in the strict sense but also to agricultural products (wines, grains, fruit, cattle, etc.) and to mineral products which are put into trade (mineral waters, etc.)."

During subsequent conferences for the revision of the Convention, this definition of industrial property was confirmed and further specified. The term "agricultural industries" was introduced in 1911 (Bureau International de L'Union, 1911:245). During the 1925 conference, the range of designated protectable subject matter was expanded to include tobacco leaf, at the request of the Cuban and Brazilian delegates (Bureau International de L'Union, 1926:517,535). Nine years later, Czechoslovakia proposed the inclusion of beer, Hungary put forward flour, and Belgium wanted to have flowers explicitly mentioned (Bureau International de L'Union, 1934:341-342). Even though the protectability of these products was not disputed in 1934, some delegates stated that the inclusion of such a rather arbitrary list could give the false impression that other agricultural product groups were excluded from protection. For this reason, the International Chamber of Commerce proposed the addition that industrial property referred "to all natural and manufactured products" (Bureau International de L'Union, 1934:342). The final text, which to this day remains unchanged, says that:

> "Industrial property shall be understood in the broadest sense and shall apply not only to industry and commerce proper, but likewise to agricultural and extractive industries and to all manufactured or natural products, for example, wines, grains, tobacco leaf, fruit, cattle, minerals, mineral waters, beer, flowers, and flour."

The broad interpretation of the term industrial property and the explicit reference to a list of agricultural and food products have recently induced authoritative patent lawyers to conclude that the original signatories to the Paris Convention "signalled their conviction that patent protection for agricultural living-matter inventions, plants and animals alike, was a desired objective" (Bent *et al.,* 1987:41). This is incorrect, however.

During the Paris Convention's preparatory meetings, as well as during the conferences on its revision, it was confirmed that the scope of the Paris Convention included agricultural industries, not in terms of patents, but rather in view of trademark and tradename protection. Some countries wished to ensure protection of the trade names of important national agricultural industries and therefore requested an explicit mention of these sectors. This is confirmed by French participants in the preparatory process. They state that the incorporation of the explanatory note on industrial property ensured that "the name or mark of a farmer, a cattle breeder , or a mineral water spring owner would be protected in the same way as was the mark or name of a factory." (translation RP/JvW) (Pouillet and Plé, 1896:14). Even when in 1934 plants were discussed for the first time, because of the Belgian proposal that flowers be included, there was no reference to patents. Not that plant varieties were excluded from patent protection in the Paris Convention, but until the 1930s, the patenting of plants was generally not a realistic option for patent lawyers.

The legal and technical obstacles to plant patenting envisaged by the lawyers played an important role in the national disputes on the issue in the 1920s and beyond. Below we will examine these disputes in the four countries where the legal protection of plants was most topical: Germany, France, The Netherlands, and the USA.

2.3.4 Practical hurdles hamper plant patents in Europe

As indicated above, during the First Agro-Food Order the 'botanical monopoly' remained the sole protection mechanism for colonial breeders to prevent their varieties from being propagated and used as source for further breeding in rival colonies. The monopoly could be combined with mutual protection and premium systems to encourage plant breeding, as has been illustrated by the sugar industry initiatives in the Dutch East Indies. In Europe, however, the breeders opted for patents, granted and enforced by the government.

The horticultural industry was the first to raise the issue of plant patents in Europe. As fruit, nut and ornamental varieties could easily be cloned, the nurseries had a great interest in legal protection against unauthorized propagation by competitors. Some resort was provided by trademark law. A trademark protects a word, such as the company name, or a logo, under which a plant variety can be sold. The usefulness of trademarks is limited, however. The variety itself is not protected and can freely be commercialized by others under another name.

The horticultural industry expected more from patent protection. The first demands in this direction were voiced in *France* during the Fruit Growing Congresses of 1904 and 1911 (Heitz, 1991:20). In 1921, Deputy Ricolfi drafted a bill to allow horticultural patents. His initiative was a reaction to a negative decision of the Commercial Tribunal in Nice, in the same year, on a case involving the unauthorized reproduction of a carnation. Ricolfi's bill did not pass, however, and neither did the bills he subsequently tabled in 1925 and 1928 (Hermitte, 1988:42).

In *the Netherlands* too, patents were not awarded to plants. When submitting the new Dutch patent bill in 1910, the government had already showed its reluctance to accept patent coverage for biological material. It indicated that in areas such as soil cultivation and cattle breeding "where the results depend less on human will than in industry", novelties could not usually be considered as inventions (translation RP/JvW) (*Memorie van Antwoord*, 10-11 May 1910, quoted in Verhulst, 1947:79). Apparently it was commonly accepted among the Dutch that plant patenting was inhibited by insufficient biological and taxonomic knowledge. In 1920 the Dutch Horticultural Council advised against the plant patent, basically because a "Botanical Patent Bureau" would not be able to examine the novelty aspect of a variety. At the end of the 1920s, reference collections of fields crops were still in their infancy, while detailed descriptions and photographs of plants in subsequent stages of their development were incomplete, and adequate identification techniques unavailable (Van Harreveld, 1934:59-62).

Legal protection for plant material was also topical in *Germany*. In order to offer breeders some support, the German Ministry of Agriculture and Nutrition prepared a Plant Breeders' Protection law in the early 1930s. The bill did not survive the *Reichsrat*, however. The proposed restrictions on the marketing of on-farm saved seed were unacceptable to farmers, and not effective enough for breeders (Van Harreveld, 1934:102; Flitner, 1995:83). At the same time, efforts were under way to have plant varieties protected by patent law. The first positive decisions were issued by the German patent office in the mid-1930s in relation to seed of tobacco, lupin, and garden pea (Heitz, 1991:28-29). Since then, "over 100 new plant varieties managed to obtain patent protection in Germany" (Lange, 1985:30-31). However, the patent route in Germany would later in the 1930s be interrupted by the Nazi regime, a development we will deal with in the next chapter.

2.3.5 The 'try-out' of legal plant protection in the USA

Like the German breeders in Europe, American breeders were also successful in their attempts to have plants patented. Similarly to Europe, the earliest calls for legal protection of plants came from the American horticultural industry, and can be traced back to the 1870s (Fowler, 1994:80). Plant patenting, however, was discouraged at the end of the 19th century. In 1889, the U.S. Commissioner of Patents upheld an earlier rejection of a patent application for a fibre identified in the needles of a pine tree. The formulation of the Commissioner, which became a "landmark doctrine", noted that ascertaining the composition of the trees in the forest was:

> "not a patentable invention, recognized by statute, any more than to find a new gem or jewel in the earth would entitle the discoverer to patent all gems which should be subsequently found." (*Ex Parte Latimer*, March 12, 1889, quoted in Bugos and Kevles 1991:5).

As a result, it became a fundamental tenet of American patent law that, in general, no protection could be obtained for products of nature, either inanimate or living (Bugos and Kevles, 1991:5).

In 1906, U.S. Congress considered a bill that originally aimed at the strengthening of plant trademarks, but had been revised to allow patents for horticultural plants, trees, and vines. Neither this bill, nor several others tabled in subsequent years, survived Congress (Fowler, 1994:81). The reluctant attitude of Congress induced the horticultural industry to create a lobby group, the National Committee on Plant Patents. Ever since, the interest of the horticultural industry in patent protection only grew, as did the U.S. market for horticultural products. In 1928, the market for ornamental plants alone was estimated at one billion dollars (Bugos and Kevles, 1991:6).

In the late 1920s, the nurseries' efforts had more success. A plant patent bill was tabled by Paul Stark, principal of the largest American horticultural breeding company, the Stark Brothers Nursery. The bill was introduced in Congress by a senator who also was one of the country's largest fruit growers, and fell in fertile ground (Bugos and Kevles, 1991:6). The industrial lobby pointed out the positive impact that plant patenting would have on the development of innovative varieties, resistant to pests and diseases. The nurserymen equated plant selectors and breeders with great American inventors, such as Thomas Edison and Henry Ford. Edison, who had been a friend of the renowned American breeder Luther Burbank, personally sent a telegram to Congress in support of plant patents (Fowler, 1994:80-84). The innovation-encouragement argument was nonsense, however. The alleged breeding of the nurseries consisted largely of propagating or crossing plants, using techniques that were well known. Practically all new varieties of fruits were the result of chance discovery of seedlings and naturally occurring bud mutations (Fowler, 1994:74,76-77).[18]

Most scientific breeding research was taking place in field crops, but these were not covered by the patent bill. The bill provided patent protection only for new asexually (vegetatively) reproducing plants, which generally encompass varieties of fruit, nuts and flowers. As most field crops, such as cereals, were not eligible for protection, the American seed industry had little to win from the plant patent bill. Nevertheless, it supported the bill as a tactical move. Stark had convinced the American Seed Trade Association (ASTA) to establish first the principle of property rights for plant breeders. Later it could be attempted to widen the scope of protection to include all seed propagated plants (Fowler, 1994:83).

Legal obstacles to plant patenting were removed by the House Committee on Patents. In a report on the bill, the Committee delivered a new interpretation of the prevailing "product-of-nature" doctrine. It stated that, while a new plant variety found in the field was a product of nature and, hence, not patentable, a new variety arising from cultivation was patentable since it was created by "human agency". The committee compared a plant originator with a chemist developing new compositions of matter. Both take the materials of nature, exploit its laws, and devise a new and useful product (Bugos and Kevles, 1991:7).

The accommodation of patent law to plants implied that several of the key requirements for protection had to be weakened or dropped. For example, rather than a worldwide novel invention, the bill specified that the new variety had to be 'distinct' from existing varieties, but it failed to define this term (Fowler, 1994:90). As a result, patents were allowed on plants, even naturally-occurring ones that were hardly different from others. As long as human intervention had been required to reproduce the plant asexually, the plant was eligible for protection. Neither did the patent bill contain the standard patent requirement of disclosing the invention sufficiently to enable someone skilled in the art to reproduce it. The bill only required that the description be as accurate as was reasonably possible (Bugos and Kevles, 1991:9-10).

The bill eventually passed Congress in 1930 without significant opposition (Fowler, 1994:87) and became known as the Plant Patent Act (PPA). The act appealed as a farmers' and plant breeders' relief measure, but did nothing for direct farm relief. Important for the Republican majority in Congress was the prospect that stimulation of private breeding activities could reduce the direct governmental involvement in plant breeding and agriculture (Doyle, 1985:51; Bugos and Kevles, 1991:7). It is doubtful whether the PPA has ever met this objective, however. The combination of a weak law, the limited scope, and the absence of a strong examination system hindered the PPA in stimulating innovation. An evaluation of the PPA, 57 years after its adoption, concluded that the act has had very little impact on private investment in fruit breeding (Stallmann and Schmid, 1987:434). Nevertheless, the adoption of the PPA was a historic event in the sense that it was the first time that plant varieties were legally protected as intellectual property.

Summary

Born out of a protectionist trade ideology, the international patent system was founded in 1883 with the conclusion of the Paris Convention, which harmonized national patent laws and enabled an extension of the territorial scope of patents to all member states. In a political environment that was generally favourable to technological innovation in view of national economic competitiveness, new plant varieties began to be appreciated as innovations.

The interest in legal protection of plant varieties during the First Agro-Food Order was most pronounced in the horticultural industry. Competition in this sector was intense, cloning of fruit and ornamental varieties was relatively simple, and purchase contract and trademarks were inadequate to curb unauthorized exploitation of plants for propagation or for producing new varieties.

Practical reasons, however, hampered the protection of plants under the patent system during the First Agro-Food Order. Taxonomic and descriptive data to categorize plants were deficient, while techniques to control the design of plants fully were not mastered. Eventually, only fruit and ornamental plants were to become patentable in Germany and the USA, and this was not until 1930.

Conclusions

During the First Agro-Food Order, the international division of labour in agriculture consisted of two different trade patterns. On the one hand, there was the traditional intra-colonial trade, based on the complementarity of products resulting from different climates, within and between the European colonial empires. On the other hand, inter*national* trade in temperate agricultural products emerged among core capitalist countries. Concern for competitiveness of the agricultural sector in their homeland instigated governments to support producers, basically by laying the foundation for a rationalization of crop development.

Governments in the then industrializing countries began their interventions by regulating the national seed market. The establishment of seed testing stations and seed certification procedures was to the advantage of farmers *and* private firms involved in plant breeding and seed production. The reduction of fraud in the seed sector raised the quality of seed, and facilitated opportunities for further capital investment in crop development. On the other hand, the seed market regulating measures also limited commercial freedom and provoked the seed industry to challenge too strict governmental control. Due to the stronger private seed industry lobby in the USA, the American seed market regulation was less strict than in most

European countries.

Governmental involvement in crop development was a major prerequisite for the re-discovery in 1900 of Mendel's insights into the heredity mechanism of living organisms. The new genetic principles offered the prospect of more effective control in the design of new plant varieties, and an expansion in the production of temperate crops. Between 1900 and the 1930s, Mendel's laws would gradually be elaborated within academic circles on both sides of the ocean. In the USA, genetics was first employed to crop development with the intention to curb surplus production of wheat through crop diversification. The new research initiatives induced the USDA to augment investments in the conservation of temperate seed and plant material. However, the application of Mendelian genetics had no *direct* effect on the competitiveness of the U.S. agricultural sector. The American efforts in research and conservation would contribute to the dominance of the American farm sector only after the First Agro-Food Order.

Governments in Europe were less keen to fund genetic research for crop development in temperate crops. Throughout the First Agro-Food Order, they continued to focus on their colonial territories, and on tropical agricultural production. The relatively slow application of Mendelian genetics in Europe was also expressed in the plant collection and conservation strategies. The European governments apparently continued to facilitate the so-called botanical exchange. They slowly turned away from the inter-colonial exchange of individual plant species to plant selection based on visible characteristics of plants.

With the expansion of capital investment in crop development, a new interest group emerged that required governmental support in the legal protection of its creations. By drawing a comparison between inventors and plant breeders, the breeding and seed-producing firms in Europe and in the USA justified their demands. The interest in legal protection of plant varieties became most pronounced in the horticultural sector. Competition in this sector was intense, and the cloning of fruit and ornamental varieties was relatively easy. Due to the lack of technical control in the biological reproduction process, and the inability to distinguish adequately among varieties, legal protection of plants would become available only in the early 1930s, in Germany and the USA.

Notes

[1] Hyams (1971:89), for example, offers an overview of the continuous infection of introduced *Coffea arabica* selections after 1650 in practically all colonies where it was introduced. Best known is the infection of coffee plantations in Ceylon around 1700 by coffee rust, which induced a change over to tea production (Monaco, 1977:59).

[2] By 1900, the French organization consisted of 84 agricultural experiment stations, a number that rose to more than 115 in 1930 (Busch *et al.*, 1995:133).

[3] Germany did not rely on intra-colonial plant transfers as much as other colonial powers, because its colonial possessions were limited. Rather, material was collected from other (rival) colonies (e.g. in Central America), or from non-colonized areas (Flitner, 1995:37).

[4] *Cinchona* seeds had earlier been introduced in Java in 1854 by the former Superintendent of the Dutch Botanic Gardens on Java, J.K. Hasskarl. One year earlier Hasskarl had entered the Caravaya region of Bolivia and Southern Peru searching for *Cinchona* plants. He travelled under an assumed name and conducted his business "in secret and through bribery" (Brockway 1988:114). Knowing the importance of Hasskarl's activities, the Dutch government sent a special naval ship to Peru to pick up Hasskarl and his bounty (Van Harreveld, 1934:135). However, all but two of Hasskarl's specimens died before they reached Java, and the seeds he brought out yielded valueless trees, so that Dutch interests in *Cinchona* were frustrated for a decade (Brockway, 1988:114).

[5] A pure line consists of a series of descending plants in which a breeder is able to maintain specific genetic characteristics after crossing.

[6] By 1880, 40 LGUs were operational (Doyle, 1985:352), while the USDA maintained about 85 experiment stations in 1930 (Busch and Sachs, 1981:135).

[7] The John Innes Horticultural Institution (established in 1910), the Plant Breeding Institute (1912), the Welsh Plant Breeding Station (1919), and the Scottish Plant Breeding Station (1921).

[8] In 1929, two-thirds of the total Dutch potato acreage was planted to Veenhuizen's varieties. Most successful has been *Bintje,* a potato variety released in 1905 and exported to many countries (Addens 1952:78,100).

[9] *Instituut voor Veredeling van Landbouwgewassen,* later called the *Instituut voor Plantenveredeling* (IVP).

[10] *Forschungsinstitut für Pflanzenzüchtung.* The institute was part of Germany's primal research organization, the *Kaiser-Wilhelm-Gesellschaft* (KWG) (Flitner, 1995:49).

[11] An illustration of the incidental (and not always effective) character of plant introduction in the Netherlands is the story of the completely unknown world traveller, the widow Mrs W. Storm-van der Chijs. Mrs Storm was greatly interested in potato production. On one of her travels that had brought her to Virginia, USA, she had observed that potatoes grown from Mexican seed-potatoes did not suffer from any major disease. She brought up this issue during the 1865 Agricultural Congress *(Landbouwhuishoudkundig Congres)* in the Netherlands. She argued that the import of this plant material could be of great value for local producers, and mentioned the address of a contact person, Mr. Goodrich, living in New York. A Dutch seed trader, Mr. Nagel, offered to write to Mr. Goodrich also on behalf of twelve interested farmers. During the subsequent Congress, a year later, Mr. Nagel reported that, to his regret, he had not received any reply to his request for planting material from the USA, even though his letter "was stamped and put in the English language." (Addens, 1952:77).

[12] The system was administered by the 'Society for granting premiums for sugar cane breeding' (*Vereeniging voor het verleenen van premies voor het kweeken van suikerrietvarieteiten*). In order to be eligible for the premium, sugar cane breeders had to register and deposit their new varieties with the experimental stations. Cane growers had to report the size of the area planted to each variety. They could use registered varieties freely for three years. For the next two years, a fee per ha was payable, and half of that amount in subsequent years (Van Harreveld, 1934:128).

[13] Sugar cane variety POJ 2878 was so successful that it was eventually grown in almost every cane producing country in the world, and it became an ancestor of every modern cane variety (Robinson, 1996:227).

[14] In fact, there was no reference to patent law or the Paris Convention of which the Dutch East Indies was a member, but to the 'Treaty on the Protection of Scientific Inventions'. This treaty, approved in the League of Nations in 1923 was submitted to all nations involved in the League in 1930.

[15] The official name of patent is "letters patent", a literal translation from the Latin *litterae patentes.* Letters patent are official documents by which certain rights, privileges, or titles are conferred. Among such open letters is the 'patent for invention', in short: patent (WIPO, 1988:19).

[16] The absence of a patent law in the Netherlands enabled the butter industry (that would later merge into

Unilever) to produce margarine after 1872 by freely using an invention by the French chemist Mège Mouriès. This invention was patented in France, Great Britain (Europe's main margarine market), and Prussia. In another industrial field, Gerard Philips in 1891 was able to start the production of incandescent lamps on the basis of Edison's technology without having to remove the obstacle of an existing patent. This was an advantage vis-à-vis foreign competitors, such as the *Allgemeine Elektrizitäts-Gesellschaft* (A.E.G.) in Germany, which had to cope with restricting licensing conditions imposed by Edison (Schiff 1971:5268). In Switzerland, the silk-dyeing factory Clavel (later CIBA) in Basel could freely start the production of an aniline dye (*"mauve"*) based on an invention of Perkin, only three years after this invention was patented in England. Also the founders of the Swiss firm Buergin & Alioth had freely used basic inventions patented abroad for building generators (Schiff, 1971:98-106).

[17] When the Paris Convention came into effect in 1884, 11 states had adhered: Belgium, Ecuador, France and its protectorate Tunisia, Great Britain, Italy, Portugal, Serbia, Spain, Switzerland, and the Netherlands. The USA, Japan and Germany would follow in 1887, 1899 and 1903 respectively (United Nations 1975, Annex 1).

[18] One Senator justified his support for the Plant Patent Bill by pointing out the years of effort necessary to develop the 'Delicious' apple variety, while in fact this apple was a naturally-occurring mutated plant, simply found in his field by a farmer who had sold it for a few thousand dollars to Stark Brothers (Fowler, 1994:74,86).

Chapter 3

The 1930s: Laying the Foundations for Industrial Crop Development

The international division of labour in agriculture that characterized the First Agro-Food Order was put under severe pressure during the economic depression of the 1930s. World market prices plummeted after 1929, both for temperate and for tropical foodstuffs, notably coffee, rubber and sugar, inducing protective measures in Europe and the colonies (Kindleberger, 1973:94-96). National food security was at stake and in a period of social instability, rural unrest had to be prevented. In those countries where protective measures already existed, such as France, Germany and Italy, national prices for agricultural products were sustained at a level up to two or three times the world market price (Tracy, 1982:130). Moreover, also in countries where a free-trade policy had prevailed, such as Great Britain, Denmark, the Netherlands, and the USA, farmers began to receive state support. Tariff and non-tariff barriers protected the national market from cheap, foreign agricultural products. The closing of the European market added to the crisis in the food exporting countries, notably the USA, which desperately tried to sell its cereals at any price abroad (Figure 3.1). Agricultural prices recovered from 1933 onwards but remained generally volatile throughout the decade.

The growing presence of the state in agriculture during the Great Depression was part of an overall economic interventionist policy by government pursued throughout the Western world. State intervention intended to save capitalist principles at a time when *laissez faire* policy had failed to revitalize the international economy. Also in agriculture, the existing, limited governmental involvement in agriculture was no longer appropriate. The economic crisis had put the main agricultural production and trade patterns under severe pressure and made alternative strategies necessary to organize agricultural production.

In most countries, governments resorted to *ad hoc* and short-term protective measures (Tracy, 1982:138). There were two exceptions: Germany and the USA. In these two countries, for the first time, a more or less coherent, long-term agricultural production strategy was developed in which the state would play a central role. The final objectives of the respective strategies, as well as the means to achieve them, differed entirely, however. The Nazi regime aimed at autarky in food production *(Nahrungsfreiheit)* in view of its envisaged territorial expansion by war, while in the USA, the Roosevelt administration intended to find a structural solution to surplus production and unstable food supply.

This chapter discusses the development of specific crop development policies against the background of emerging new agricultural production strategies in both Germany and the USA in the 1930s. Both in Germany and the USA the government became actively involved in plant breeding, plant collection and conservation, and the protection of plant breeders. These policies would set the trend for crop development policies followed by most Western countries during the Second Agro-Food Order.

We start this chapter, however, with an examination of crop development policy in the Soviet Union during the *interbellum*. Russian scientists, who were educated in Mendelian genetics, would leave their marks on Soviet plant breeding and conservation policies until their scientific paradigm was considered to contradict communist principles. While their scientific insights were neglected in the Soviet Union itself, they proved to be crucial to the plant conservation policy of Germany and the USA and other Western countries during the Second Agro-Food Order.

Figure 3.1 Wheat and rye exports by main producing countries (1924-1938) (in million tonnes)

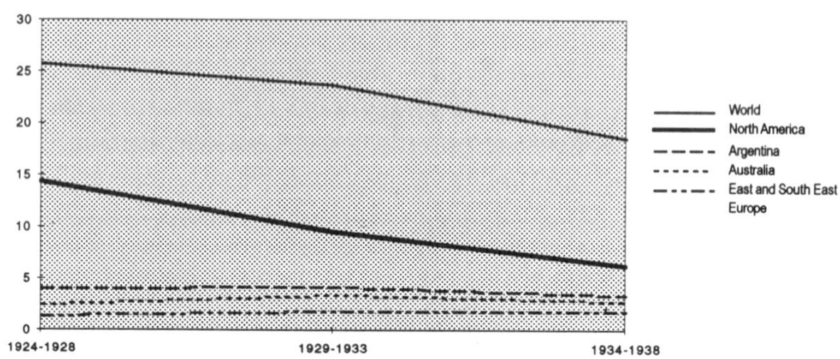

(Data derived from Tracy, 1982:143)

3.1 Mendelian genetics delinked from Soviet crop development

Like other wheat exporting countries in the world, the Soviet Union, in the early 1930s, was hit by plummeting world market prices (Kindleberger 1973:93)[1]. The problem was aggravated by crop failures in 1932 and 1933. The wheat crisis ignited a heavy scientific and political controversy on the position of Mendelian genetics as the prevailing scientific paradigm in crop development. A split occurred between those scientists who had been educated in Germany and Great Britain on Mendelian genetics, and those who had not. According to the latter, the food crisis proved that Mendelian genetics had failed - a position supported by the Communist Party. The 'Mendelians' however argued that a boost in yields would require further elaboration of their theories and experiments. The outcome of this discussion would have a decisive impact on Soviet crop development policies.

Plant breeding activities had started already in the pre-Soviet era under the coordination of the Bureau of Applied Botany (established in St. Petersburg in 1894). Since 1930 the Bureau is known as the All-Union Institute of Plant Industry (in Russian initials V.I.R.). The VIR formed the nucleus of the All-Union Academy of Agricultural Sciences. The VIR was the major crop development institute in the Soviet Union. It had plant breeding programmes and maintained a national network of at least 40 satellite seedbanks and breeding stations[2]. Special teams of taxonomists and botanists travelled around the world to extend the collections. The VIR seed collections in the 1930s contained the le-gendary number of 250,000 samples, collected from over 50 countries (Plucknett *et al.*, 1987:162). Since then, only the U.S. National Seed Storage Laboratory (NSSL, established in 1957, see section 4.4.1) has contained a collection of that magnitude (James, 1961:322).

Soviet advances in genetics and taxonomy in the 1920s and 1930s had a great influence on the understanding of the value of landraces and wild relatives of cultivars for crop development. Key figure was Nicolai Ivanovich Vavilov, who formed the bridge between genetics and crop development in the Soviet Union. Vavilov received his education in genetics in the 1910s at Cambridge University from the world's leading geneticists. Almost immediately after he had finished his studies he became the new director of the All-Union Institute of Applied Botany and New Crops (the forerunner of the VIR) and the Institute of Genetics of the Academy of Sciences. These functions allowed him to combine his knowledge of Mendelian genetics with a great botanical interest, which went far beyond the borders of the Union. In 1926 his many botanical expeditions abroad resulted in a theory on the "Universal centers of origin of the most important cultivated plants" (Vavilov, 1992)[3], which held that the original 'homeland' of cultivated plants, as well as early forms of agriculture and human culture, were to be found in only seven specific geographical regions in the world (Figure 3.2).

Vavilov's many travels to practically all continents, his frequent contacts with the most prominent geneticists of his time, his massive production of seminal books and publications, and his managerial and linguistic qualities contributed greatly to his international fame. Because of Vavilov's work, Western agricultural scientists could greatly benefit from Soviet agricultural research. The extensive VIR botanical collection expeditions served as a guiding example for German and American scientists. In spite of the political barriers, plant material was continuously exchanged between the VIR and Western research institutes, at least until the late 1930s (Hyland, 1984:12). In the Soviet Union itself, however, Vavilov's ideas, contacts and work were not appreciated. Despite his stature as the Communist party's chief biologist, Vavilov's belief in the importance of genetics for agriculture

Figure 3.2 Centres of origin of the most important crops according to Vavilov (1929)

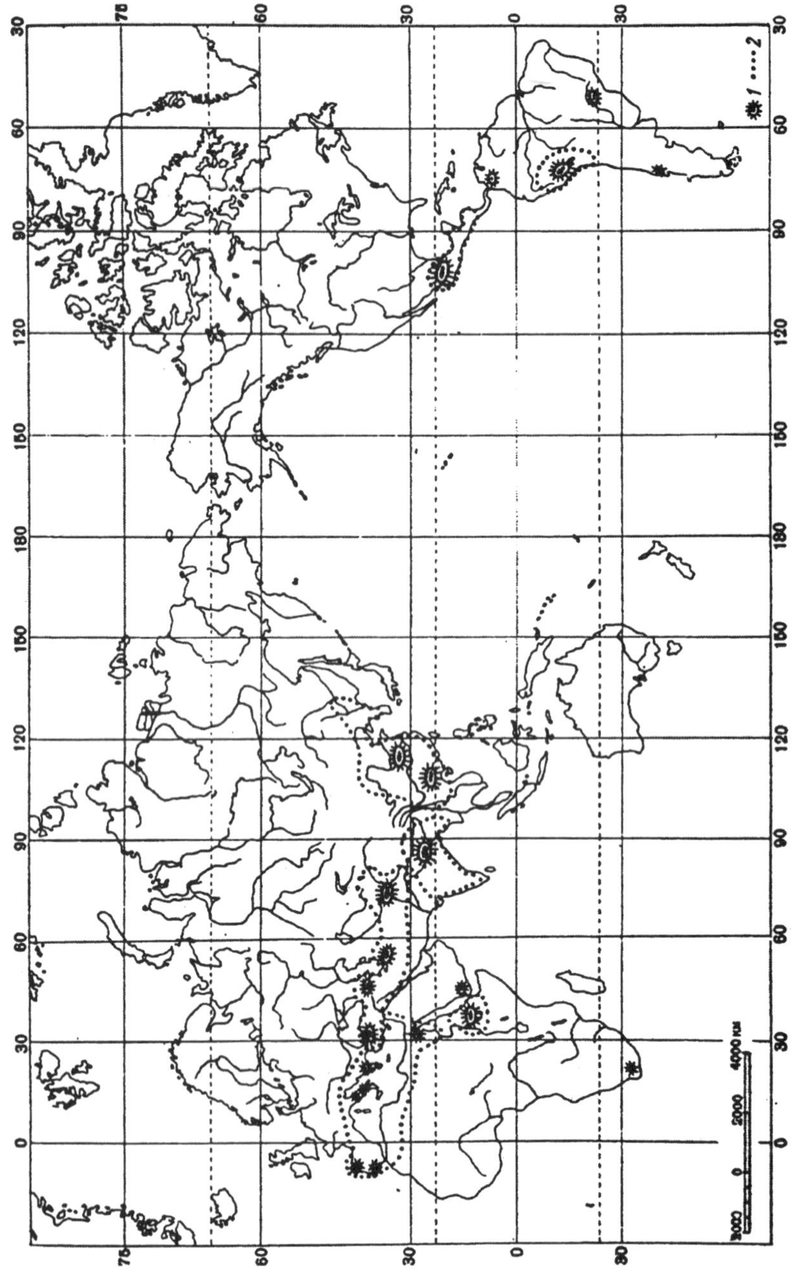

(Source: Vavilov, 1992)

brought him into serious conflict with the regime. Stalin's agricultural industrialization programme did not permit any dependence on basic Western science, notably genetics.

The communist agricultural policy of combining collectivization with the introduction of monocultures and mechanization, demanded cereal crops (especially wheat) which could be grown under the most adverse circumstances in the most remote corners of the Union. Stress-tolerant varieties were required that could thrive in the harsher climatic zones in the steppe fields, east of the Volga. Initially, the Soviet policy proved to be relatively successful, so that Stalin was able to claim that State farms would soon cover "as large an area under grain as the whole of Argentina [and] one million hectares more ... than ... Canada has to-day." (Dobb, 1972:225-226). In reality, the increases in Soviet food production were far behind the projections. The food crises of 1932-1933 led to a political controversy on the merits of Mendelian genetics and of its Soviet protagonist, Vavilov.

Mendelianism was questioned by members of the Soviet Academy of Science, notably by T.D. Lysenko, a personal friend of Stalin, and I.I. Prezent. The battle that emerged between them and Vavilov culminated at the 1936 Genetical Congress. Lysenko and Prezent declared that the early work of genetics, especially that of Bateson's school (see section 2.2.4), was subversive to Darwin's evolution theory and, in fact, a product of "clerical reaction against evolutionary biology." (Hudson and Richens, 1946:70). In the eyes of Lysenko and Prezent, Darwinism was much more in line with marxist-dialectical interpretations of how plants (and other living organisms) adapt to changing environmental conditions (Hudson and Richens, 1946:18-20). Mendel's laws suggest that the adaptation of a plant to its environment was preordained by a genetic 'system' that could, or could not, express adaptational characteristics. It obliged breeders to sort out the desired genetic characteristics of a plant (e.g. the ability to grow on infertile tundra soil) by means of a long and complex system of crossing and selecting plants. This plant breeding method made Mendelian genetics, in the eyes of Lysenko and his comrades too complex, costly, and dangerously close to fascist ideas on genetic 'pureness'. In short, Mendelian genetics violated the public interest.

Lysenko and Prezent were convinced they could develop a much simpler and a more effective crop development strategy based on Darwin's observation that organisms are constantly adapting themselves to changing environments. In this theory, a plant subjected to a changing environment would, in order to survive, automatically change its internal genetic constitution within one or a few generations. Translated into breeding practice, this phenomenon, which was referred to as "assimilation", would have a tremendous advantage over Mendelian plant breeding (Hudson and Richens, 1946:18). It implied that a breeder could avoid a painstaking selection process to genetically adapt plants by simply exposing plants to adverse circumstances and waiting for their adaptive response. Lysenko and Prezent almost immediately gained wide political support for their ideas, especially when they claimed assimilation breeding could well form the basis for a steep increase in agricultural production (Laird and Laird, 1970:51).

The Soviet rejection of Mendelian genetics reached a peak during the 7th International Genetical Congress in Edinburgh in 1939. Vavilov was appointed to chair the meeting, yet his supervisors in Moscow did not allow him to attend the congress (Dixon, 1993).[4] A little later Vavilov was arrested because of alleged anti-Soviet activities.[5] He died in prison in 1943 in unknown circumstances (Vavilov, 1997). Vavilov's death signalled not only the final delinkage of Soviet agricultural science from its Western counterpart, but also of Mendelian genetics from crop development. Under Khrushchev's leadership, after the Second World War, 'Lysenkoism' would be used to obtain the communist party's support for what later turned out to be over-ambitious plans to boost agricultural production.[6]

3.2 Nazi German agricultural autarky

When the National Socialists came to power in 1933, they had a well-elaborated agricultural policy designed by Walther Darré, Hitler's first National Socialist Minister of Agriculture. Capitalizing on the rural dissatisfaction in Germany, the Nazis offered farmers the prospect of relief from the high debt burden, taxes, and decreasing agricultural prices. This policy was embedded in the ***Blut und Boden*** ideology, which considered the German farmers descendants of Germanic settler tribes, with a close relationship to the soil. The farmer was therefore not only the source of food, but also the "blood source" of the country (Darré quoted in Tracy, 1982:202). Initially, a combination of measures did improve subsistence in the countryside, but after 1935 agriculture was downgraded in favour of the armament programme at the expense of the farmer's income (Koning, 1986:118-119). Agricultural production and trade were put under total state control by the National Socialist Food Corporation (in German initials RNS), while agricultural imports were drastically reduced, except for maize, because of the shortage of home-grown fodder.

The Nazi regime, more seriously than previous governments, also tackled the problem of fragmentation of the land. To this end the *"Erbhofgesetz"*[7] was designed, a racial farm legislation that included elements of a structural agricultural production strategy. For the first time, a law indicated the minimum size of a potentially viable farm: farms had to be large enough to support a family (Tracy, 1982:207). Farms that could not meet this requirement could not be registered under the law and were doomed to disappear. One of the intentions of the legislation was to reduce the number of people whose income depended on agricultural production. As shown below, this policy was also an essential element of the American agricultural policy under the New Deal. After 1945, the distinction between 'viable' and 'non-viable' farms would become the cornerstone for agricultural policy in all industrialized countries.

3.2.1 Conservation as a scientific-military operation

As in the Soviet Union, the German interest in developing a crop development policy formed an essential part of the long-term agricultural production strategy. During the First World War, German agro industry had lost most of its capacity to produce seed. This loss, and the memory of the "Hungerblockade" by the English navy in 1916-1917, fed National Socialist ideas for a strong and new autarkic agricultural sector. Credos such as accomplishing "Food Liberty" *(Nahrungsfreiheit)* and the closing of the "protein and fat gap" *(Eiweiss und Fettlücke)* became part of the National Socialist programme (Flitner, 1995:68-74). In 1933, 3.6 billion *Reichsmark* (RM) had to be spent on food imports, mainly fats. After a poor harvest in 1934-35 the agricultural programme was incorporated in Göring's Four Year Plan (1936), which was designed to prepare the country for war. Indeed, Germany entered the war with a healthy agricultural sector and a grain reserve of some 6 to 7 million tonnes (Cecil, 1979:29).

Achieving self-sufficiency in food and fodder also became the central aim of the crop development policy of the Nazi regime. The regulation of the seed market, the support for plant breeders, the interest in collecting and conserving genetic resources of plants, all this served the function of reducing the dependence on food imports. The quest for food autarky tied political, scientific and military elements together, a tendency which was probably best expressed in the search for seed and plants. The expeditions of Nazi taxonomists evolved into scientific-military operations in which occupied territory was effectively screened for new seed.

The interest in botany and crop development materialized in the establishment of the *Kaiser-Wilhelm-Institut* (KWI), in Müncheberg, near Berlin. During the war, the KWI rapidly became the key scientific institute for agricultural research in Germany. In 1944, the KWI housed six sections, with a division very similar to today's research centres: genetics, cytology, systematics, physiology, cultivars, and horticulture. Extensive financial support was offered by the chemical and fertilizer industry and the banking sector, allowing the Institute's annual budget to rise from 120,000 RM in 1930 to 1.5 million RM in 1939 (Flitner, 1995:87). These investments helped Germany to regain its leading position in Mendelian genetics, which had been seriously affected after the First World War.

Extensive genetic resources collection activities became part of the German scientific programmes to reach food autarky. One of the most significant initiatives of the KWI in this field was the 10 months trial to the Hindu Kush mountains (predominantly Northeast Afghanistan), financed by the German scientific society named *Deutsche Forschungsgemeinschaft* in 1935 (Flitner, 1995:74-86).[8] Military successes of the German army during the Second World War allowed the KWI to extend its crop development and collection efforts to occupied territories in Central and Eastern Europe, particularly the Balkans. Direct military or paramilitary backup was received for plant collecting expeditions in Spain during the Civil War, and in Abyssinia (Ethiopia) in cooperation with the Italian army. In spite of (or perhaps due to) the close contacts between the VIR under Vavilov and the KWI in the 1920s and 1930s (Loskutov, forthcoming), a special SS-unit for Collection in Russia (the *"Russland-Sammelkommando"*) was established, to take charge of the "systematic robbery" of the seed collections (Flitner, 1995:113-121). Many collections were captured, with the exception of the large collection at the VIR in Leningrad. Some of the collections that arrived in Germany were lost at the end of the war, some fell into British, and some into Soviet hands. A substantial part was brought over to Gatersleben, where it formed the basis of the East German seedbank after 1946 (Hammer, 1993:1).

3.2.2 Breeders benefit from authoritarian seed market control

The power shift in 1933 in favour of the National Socialists brought about a radical change in the organization of plant breeding and the seed market. Initiatives to establish a Plant Breeders' Protection Law, and the tendency to protect plants under patent law, were set back for political and technical reasons.

Political opposition to plant patenting came especially from the RNS which was "decidedly anti-patent" (Bent *et al.*, 1987:45). The RNS considered plant patent monopolies detrimental for plant breeding and therefore also for the Nazi agricultural production strategy. This opposition left breeders little other option than to withdraw their patent applications filed in Germany, although they maintained their applications abroad (Lange, 1985:31).

Apart from the political opposition, plant patents proved to be impracticable because it was difficult to meet the requirements of patent protection (Lange, 1985:31). A major obstacle was that breeders had to show that their 'invented' plant was applicable in industry. Also the requirement that the invention had to be replicable was too demanding at that time. Before cloning techniques became available in the 1970s, the exact replication of sexually reproducing plants was impossible. Furthermore, a patent is granted on an 'invention', but inventiveness was generally not involved in plant breeding.

The National Socialist regime came up with an alternative crop development policy, which involved the complete restructuring of the German seed sector. The main legal instrument was the Ordinance on Seed Material, promulgated in 1934, putting seed production and trade under strict state control. In order to raise agricultural production the RNS, in its own words, intended to "protect German farmers ...[from]... inferior, contaminated, hereditary-diseased seed material" (translation RP/JvW) (Flitner, 1995:81). The Ordinance, therefore, prevented the use of on-farm saved seed, and obliged farmers to purchase approved seed from breeders only. In addition, the Ordinance offered a solution for the 'varietal jumble' which, according to the agricultural establishment, had already disturbed the seed market for too long. Around 1935, the registration of plant varieties in a national variety list became obligatory. The national list was a useful tool to start a drastic, so-called, "varietal cleaning" operation removing 90% of the plant varieties from on the market (Flitner, 1995:82).[9]

According to Flitner (1995:85,130-131), the restructuring of the national seed market by the Nazi regime supported the German plant breeders, even though private exploitation rights were not granted. By obliging farmers to purchase breeders' seed, and by reducing the number of marketable varieties, the 1934 Ordinance offered private breeders a good opportunity to eliminate unauthorized reproduction of their varieties.

3.3 Stabilization of food production under the American New Deal

The Great Depression hit not only farmers in Europe, but also their main competitors on the other side of the ocean. Burdened by the low food prices of the early 1930s, the American farming community called for active government support. The political elite thought it wise to take this call very seriously, as farmers were important allies against the threat of emerging socialism.[10] This was the situation when the USA witnessed a shift of power from Republicans to Democrats in 1933, the same year in which the National Socialists took over government in Germany. The Democrats would develop a production strategy for agriculture with the intention of strengthening the linkage between agriculture and industry, and to ease social unrest in the countryside.

The Roosevelt Administration presented a series of policies for restructuring and reviving the economy, known as the 'New Deal'. The programme of the Democrats was historically unique, and contrasted with contemporary policies in Germany and the USSR in that it envisaged a central role of the state in organizing the national economy without violating the basic principles of the market economy. The new role assumed by the U.S. government was clearly expressed in agriculture.

The New Deal contained an important agricultural element, composed of a series of interrelated programmes, initiated over a number of years. These programmes, of which the Agricultural Adjustment Act (AAA) is the best known, were often contradictory as they reflected the different social-economic interests involved in agriculture. Nevertheless, the Roosevelt Administration succeeded in developing a more or less coherent strategy for tackling the problems in agriculture. The New Deal policies created the foundation for the agricultural production strategy of the Second Agro-Food Order, not only in the USA[11] but also, as we will see in chapter four, in the entire Western world. Basically, the New Deal strategy encompassed the stabilization of food supplies by establishing food reserves and offering income guarantees for farmers. This raised the purchasing power of farmers, enabling them to undertake investments and subsequently raise farm productivity. To prevent a boost in production and a corresponding collapse of agricultural prices, it was considered crucial that the minimum income standard was only applied to the wealthier ('viable'), middle-sized and large farms.[12] Non-viable farms, in fact, were given up.

The implementation of this aspect of agricultural production strategy under the New Deal was greatly supported by the events of the time. An expanding industry, the mis-croppings in the mid-1930s, and the outbreak of the Second World War, encouraged or forced a large number of share croppers and tenants to find jobs in an urban industrial environment, or in the U.S. army (Saloutos, 1982:266).

The new production strategy succeeded in raising agricultural productivity by specific crop development policies, designed by the USDA under guidance of the U.S. Secretary of Agriculture, Henry A. Wallace. Traditionally, agricultural research in the USA had been directed at production increase through area extension, rather than yield increase per hectare. But when marginal land began to be cultivated, spending on fertilization and irrigation rose. Labour costs rose as well, because the low population density in the agricultural areas could only partly be compensated by mechanization (Cochrane, 1979:185). Under these circumstances an increase of yield per hectare came within sight as a means to cut down input and labour costs. This induced the Democrats during the New Deal drastically to reverse reductions in public research expenditure, a trend set by the Republicans. The aim was to raise productivity in American agriculture by creating new, high yielding varieties.

The rising outlays for crop development were controversial among farmers. Pointing to the surplus production and low agricultural prices at the time, they questioned the benefit of research directed at higher production, rather than higher prices. Secretary Wallace, however, held the view that not *less* but *more* science would be the remedy for the agricultural crisis (Kloppenburg, 1988a:85-87).

A financial boost for agricultural research, basically through the 1935 Bankhead-Jones Act, was not the only characteristic of the New Deal crop development policy. Also significant was the type of research that was supported. Wallace had his roots in the seed industry. He was the founder of the Pioneer Hi-Bred Corn Company (which would develop into one of the world's largest seed companies), and a strong advocate of the 'pure-line' plant breeding approach. Wallace appointed a Secretary's Committee on Genetics, the results of which were reported in the 1936 and 1937 Yearbooks of Agriculture. Kloppenburg (1988a:88) argues that these reports marked the coming of age of 'agricultural genetics'. The reports focused on Mendelian genetics for the understanding of practical problems in crop development, and considered hybridization as the main breeding tool.

3.3.1 Conservation becomes the cornerstone of crop development

The crop development policy of the New Deal also increased support for conservation of landraces and wild relatives. The support was expressed in the Bankhead-Jones Act of 1935, which called for an expansion of the U.S. conservation and collection capacity. The rising interest in plant collection in the 1930s resulted from advances in Mendelian genetics. The insight that landraces and wild relatives should be assessed on genetic criteria, rather than outward characteristics, greatly expanded the number of plants that were potentially useful. Two problems had to be addressed, however: the loss of landraces because of their replacement by newly developed varieties, and the limited seed storage capacity in the USA.

The growing value attributed to genetic information in crop development in the 1930s provoked concern about the speed at which 'modern' varieties were rapidly replacing 'traditional' landraces. Plant breeders began to realize that genetic information contained in landraces would be valuable for breeders in the future. The danger of a deadlock in breeding was first realized by J. Harlan and M.L. Martini. In their seminal article on the rapid replacement of old barley landraces of 1936, they summarized the problem as follows:

> "In the hinterlands of Asia there were probably barley fields when man was young. The progenies of these fields with all their surviving variations constitute the world's priceless reservoir of germ plasm. It has waited through long centuries. Unfortunately, from the breeder's standpoint, it is now being imperiled. When new barleys replace those grown by the farmers of Ethiopia or Tibet, the world will have lost something irreplaceable." (Harlan and Martini, 1936:303).

Harlan and Martini not only revealed their concern for the loss of a 'common good', but also their awareness of the long-term interests of the plant breeding industry.

> "The plant breeder has every reason to feel gratified and undoubtedly the time is not too far distant when the entire acreage will be planted to pure-line varieties. There is, however, one rather disconcerting problem raised by the plant breeder's success. In a way, we lose whenever we gain ... The breeder is helpless without living material of diverse character." (Harlan and Martini, 1936:315-316).

Very likely this was the first time plant breeders themselves publicly expressed the paradox of the success of the spread of improved, genetically more uniform crops. Harlan and Martini were frequent travellers and had personally witnessed the gradual replacement of traditional landraces by new varieties. Their global scope on the "irreplaceable" loss of diverse germplasm, however, also reflects an American bias. Knowing that the USA traditionally is a large 'net importer' of landraces and wild relatives of food crops, the authors realized that the American breeding industry would be one of the first to suffer from the replacement of traditional plants. The fact that Harlan and Martini published their article in the same 1936 Yearbook of Agriculture in which Wallace reported on his New Deal crop development policy, contributed to the political impact of their call for more conservation capacity.

While the necessity to collect landraces and wild relatives was increasingly emphasized, the capacity to store and conserve the seed collected in remote areas of the world reached it limits. During the 1920s and early 1930s, American collection efforts were at a peak. "Hundreds of collections were brought together at the nursery centres throughout the country for screening ... [and] ... contributed more to crop improvement and production than any other time of U.S. agricultural history" (Hyland, 1984:10). (Until in the mid-1930s political conditions prevented joint conservation efforts, the USDA was in frequent contact with Vavilov and his staff (Hyland, 1984:12).) The call for more storage capacity was voiced in the late 1930s by the U.S. National Academy of Sciences Committee on Plant and Animal Stocks. The committee expressed its concern over the high loss-rate of the U.S. seed collections due to inadequate storage facilities. It would not be until after the Second World War that the USDA was allowed to build four new Plant Introduction Stations with extensive conservation facilities (Busch *et al.*, 1995:93) (see section 4.4.1).

3.3.2 Hybrids as an alternative protection paradigm

The New Deal encouraged the breeding of plant varieties that would raise productivity of agriculture through various means, although not by enacting specific intellectual property rights legislation. Of course, the USA was the first country that had established a specific plant patent system, but that was not related to the New Deal policy for agriculture. The Plant Patent Act (PPA) of 1930 had been established before the Roosevelt Administration entered into office and was specifically designed to protect asexually reproducing varieties: fruit and nut trees and ornamental plants. For the American seedsmen it had little to offer.

The absence of legal protection did not seem to be a big problem for the seed industry, however. While German seed firms were demanding - and were given - state protection against unauthorized reproduction of their varieties in the 1930s, legal protection was not a big issue for the American seed industry (Bugos and Kevles, 1991:11). American plant breeding in field crops was predominantly carried out by state-funded institutions that did not seek legal protection for their innovations. The existence of 'traditional' protective measures may have played a role too. Breeders still enjoyed some protection from seed certification, purchase contracts, and trademarks. Most important, however, was the coming of age of agricultural genetics, which made available hybrid plant varieties that rendered legal protection superfluous.

By developing hybrids, breeders took advantage of a natural phenomenon, called 'heterosis' or 'hybrid vigour', which is the tendency for offspring of genetically diverse plants to perform better than their parents. By deliberate crossing two different parental lines which contain specific characteristics, the first filial generation (F1) of this cross can express these desired characteristics along with the hybrid vigour. Yield potential is very important. In some cases, the F1 hybrids yield significantly more than 'open-pollinated varieties' (OPVs).[13]

Hybrids offer built-in protection against propagation. Unlike OPVs, the second generation of which gives yields similar to the first, the second generation of hybrid varieties is highly variable and significantly lower in yield than that of the first generation. This phenomenon has two consequences. When using hybrid varieties, farmers have to buy fresh, first generation seed every year in order to maintain the high yields, and other desirable characteristics of the hybrid varieties. Additionally, the lower yields and high variability of the second-generation seed eliminate the incentive of farmers and professional seed producers, to re-use seed from a hybrid. Only the breeder who holds both parental lines is able to produce economically viable hybrid seeds. In other words: hybrid varieties effectively prevent the double use value of plants; hybrids produce grain or fruit rather than useful seed.

This unique property of hybrids was very clearly realized right from the outset of their development. Two of the early promoters of hybrid varieties, Edward East and Donald Jones, in their classic book *Inbreeding and Outbreeding* (1919) explained the advantages of hybrids for commercial breeders.

> "It is not a method that will interest most farmers, but is something that may easily be taken up by seedsmen; in fact, it is the first time in agricultural history that a seedsman is enabled to gain full benefit from a desirable origination of his own or something that he has purchased. The man who originates devices to open our boxes of shoe polish or to autograph our camera negatives, is able to patent his product and gain the full reward for his inventiveness. The man who originates a new plant which may be of incalculable value benefit to the whole country gets nothing - not even fame - for his pains, as plants can be propagated by anyone. There is correspondingly less incentive for the production of improved types. The utilization of first generation hybrids enables the originator to keep the parental types and give out only the crossed seeds, which are less valuable for continued propagation."
> (East and Jones quoted in Bugos and Kevles, 1991:12)

For economic and technical reasons, maize was the crop for which hybrid varieties were developed first. By the end of the 19th century, livestock experts favoured the improvement of especially maize, which was the prime feed crop in the USA (Crabb, 1947:15), while the physical structure of the maize plant facilitates control of the pollination process, which is the key to inbreeding and hybridization. Moreover, the chromosome structure of maize turned out to be favourable for cytogenetic studies (Mayr, 1982:767). The first commercial maize hybrids were marketed in the 1920s, but their breakthrough came two decades later. In 1934, only 0.4% of the total American maize area was planted to hybrids; by the 1940s the percentage had risen to 60% (Dowswell *et al.*, 1996:55).

The seed industry welcomed the potential of the built-in protection in hybrid varieties. The initiator of the Plant Patent Act, Stark, explained later that, when Congress was considering the PPA in 1930, maize breeders thought they might get better and quicker protection from hybrid maize than from plant patents. Their optimism spread to other breeders who expected that hybrid varieties would also be developed in other seed crops (Bugos and Kevles, 1991:12).

The positive expectations of the inbred-hybrid breeding method were strengthened by very strong political support of this breeding strategy by Secretary of Agriculture Wallace. Wallace had conducted inbreeding experiments himself, and his company, the Pioneer Hi-Bred Corn Company, had been the first to grow commercial hybrid maize in Iowa in the 1920s (Crabb, 1947:109,152). It was also Wallace who had been instrumental in a drastic change of breeding policy within the maize section of the USDA in 1922. He advised his father, the then Secretary of Agriculture, to replace the principal maize agronomist at the USDA, who opposed the inbreeding technique, by a declared supporter (Crabb, 1947:99). This shift in personnel marked the beginning of a massive 15-year hybrid maize programme, subsidized and coordinated by the USDA (Berlan, 1997:8).

The inbreeding-hybridization route was extremely costly. Inbreeding in maize depresses yields, while heterosis occurs in only a few combinations of inbred lines. To find the right combinations, hundreds of thousands of plants had to be tested. This job could only be carried out through an "unprecedented venture" in which the brightest research leaders in the USA were investigating one of the ten gene linkage groups then known in maize (Kloppenburg, 1988a:103). The programme made available hybrid varieties of maize that marked the beginning of an ongoing period of continuous yield increase. Pointing to the maize yields, Wallace would later compare the power of heterosis with that of atomic energy by drawing an analogy between the hybrid maize programme and the 'Manhattan project' (Berlan, 1997:8; Kloppenburg, 1988a:91).

The choice of the inbreeding-hybridization technique has proved to be very influential in plant breeding. The American funds devoted to the hybridization programme supported the formation of a new approach in breeding research, at the expense of others – and this is what makes the inbreeding-hybridization technique controversial. It has been pointed out by Kloppenburg (1988a:101) and Lewontin and Berlan (1990:616-617) that the choice of this breeding strategy was political rather than scientific. The real aim would have been the abandonment of on-farm seed production. To this end, the development of cheaper alternative breeding methods was deliberately put aside. One such alternative was 'recurrent selection', a plant population improvement method which involves the self-fertilization of plants, the selection of those lines for desired characteristics, and finally the intercrossing of these superior lines. This method also involves inbreeding and heterosis, and may also result in hybrid varieties. However, this type of hybrid can be replanted by farmers and does not help the breeder to increase his return on investment.

Berlan (1997:20), and Lewontin and Berlan (1990:626) argue that the higher yield of hybrids was based on a self-fulfilling prophecy. In the mid-1930s, American farmers began to shift over to maize hybrids because they out-yielded open-pollinating maize varieties. But since hybrid development was made the only breeding strategy in maize, the open-pollinating varieties had not undergone any improvement for several years. The yield performance of hybrids was thus compared to that of maize plants that were in the same genetic state as they had been in the 1910s, when plant improvement was still directed towards aesthetic characteristics rather than to yield. In short, if alternative plant breeding techniques had been developed concurrently with the inbred-hybrid method, the yield advantage of hybrids may have been less significant.

Conclusions

During the 1930s, the state started to intervene in agriculture in all of the then industrializing countries, but most forcefully in the USA and Germany. These countries were unique, in the sense that they gradually developed a long-term and state-led strategy for a structural transformation of agricultural production. The strategies entailed specific policies on plant breeding, plant collection and conservation, and the protection of private plant breeding firms.

The spread of Mendelian genetics in the first decade of the 20th century provoked breakthroughs both in breeding and conservation. As science disclosed how genes control specific traits in plants, the search for useful genes from a broad range of plants intensified. The new genetic knowledge greatly expanded the scope of plant material with potential economic value. Not only cultivated varieties, but also landraces, and to some extent wild relatives were increasingly seen as 'resources' for breeding. This development opened the door for a new type of scientific activity of which the Russian geneticist Vavilov was the exponent: the worldwide collection of plants as germplasm resources in their centres of origin.

The strategic value of germplasm of temperate crop plants was realized by the main economic and military powers of the time: the Soviet Union, Germany and the USA. During the *interbellum* they became the key players in the worldwide collection and conservation of plant germplasm. However, geopolitical developments would lead to the Soviet Union and Germany being left behind. The Soviet communist party caused the Soviet adherents of Mendelian genetics to play down their ideas as well as their Western contacts. The outcome of the Second World War terminated the political drive in Germany to establish agricultural autarky and a strictly national germplasm collection.

The newly emerging crop development policies in Germany and the USA did not entail further intellectual property protection for plant breeders. The Nazi regime terminated the tendency of plant patenting in Germany. Instead, German breeders enjoyed *de facto* protection from an authoritarian control of the seed market in which the marketing of seed other than that of plant breeders was not allowed. In the USA, the coverage of the Plant Patent Act would remain limited basically to fruit and ornamental plants. The production of hybrid varieties was strongly developed by the Democratic government as a new protection paradigm for plant breeders.

Notes

1 Grain exports made up 36% of Russia's export in the 1910s and 1920s (Kindleberger, 1973:93).

2 The exact number is not known. Some sources spoke of over 400 seedbanks and breeding stations, but this number is contested by Frankel, a contemporary of Vavilov (personal communication Otto Frankel, Australian Commonwealth Scientific and Industrial Research Organization, CSIRO, Canberra, Australia, 1994).

[3] Vavilov's ideas were first published in 1926 as Studies in the Origin of Cultivated Plants (Vavilov, 1926). The study became widely available only in 1951 when it was translated and republished as The Origin, Variation, Immunity and Breeding of Cultivated Plants: Selected Writings of N.I. Vavilov (Vavilov, 1951).

[4] Vavilov was increasingly put under pressure. In a desperate attempt to defend himself, he managed to meet Stalin on 20 November 1939. This is how the meeting is described by Vavilov's colleague at the VIR, E.S. Yakushevsky: "... [H]e (Vavilov), after several attempts, reached his goal and was granted admission to Stalin at 10 p.m. After he arrived, he spent two hours in the reception room and at long last, after midnight, he was allowed in. Vavilov entered and greeted Stalin: "Good evening, Iosif Vissarionovich", and bowed. (Later he was told that Stalin hated being called by his first name and patronymic, but preferred being addressed as Comrade Stalin). Stalin did not answer the greetings and said: "So you are the Vavilov who fiddles with flowers, leaves, grafts and other botanical nonsense instead of helping agriculture like Academician Lysenko, Trofim Denisovich". Vavilov was not invited to sit down. Stalin walked up and down the office and left Vavilov standing. In the first instance Vavilov was taken aback, but then regained his spirit and read him a lecture about the problems of the Institute, the assemblage of the VIR seed collection, and ... everything that was done for the development of plant science, and for the development of plant breeding and seed production in this country ... After Vavilov had told all this, he calmed down, but noticed that it was like talking to a stone wall. After an hour, Stalin cut him short, saying: "You are free to go now, Mr. Vavilov". Vavilov bowed and left." (Loskutov, forthcoming)

[5] The verdict against Vavilov of 5 July 1941 was formulated as follows: "Preliminary and fullscale investigations have determined that since 1925, Vavilov has been one of the leaders of an antiSoviet organization called Peasants' Party of Labour. Since 1930 he has actively participated in a rightwing antiSoviet organization that functioned within the People's Commissariat of Agriculture of the USSR and several scientific institutions of the USSR... Pursuing interests of antiSoviet organizations, he performed acts of sabotage on a wide scale aimed at the undermining and liquidation of the collective farm system, and at the breakdown and decline of Socialist agriculture in the USSR..." (Loskutov, forthcoming)

[6] The incorrectness of Lysenkoism was indirectly admitted when Vavilov's work was officially reassessed by the Communist Party in 1955.

[7] The *Erbhofgesetz* was a legislative elaboration of the Nazi ideology and provided that farms, once registered as *Erbhof*, literally translated as 'entailed farm', could not be disposed of by the current owner and had to pass intact to a single heir. In 1938, half of the agricultural area was registered under the law (Tracy, 1982:207).

[8] Flitner (1995) points out that the Hindu Kush expedition was not only organized for agricultural reasons. The search for 'pure' seeds also served as an opportunity to test pseudo-scientific eugenic hypotheses on the "pureness" of the Aryan race. The criterion "pureness" was not only applied to the more than four thousand landraces and wild relatives collected (mostly wheat, barley and legumes), but also to the physiological characteristics of several hundred local inhabitants. It was assumed that both plants and men had hardly been exposed to racial blending, leading to what in Nazi ideology was called "softening" *(Verweichlichung)*, because of their long-term geographical and cultural isolation (Flitner 1995:74-80). How close plant breeding and eugenetics were related is illustrated by the following quote from the dissertation (1934) of Herman Kuckuck of the KWI: "An application of plant improvement methods to improve the human race cannot be carried out right away. These methods will nevertheless increase our insight and understanding of the necessity to take strong measures to preserve the human race, as already convincingly initiated by the sterilization law". (translation RP/JvW) (Kuckuck quoted in Flitner, 1995:114).

[9] The drastic reduction of varieties was undoubtedly also related to the relatively late introduction of the

national variety list in Germany. In the Netherlands, a comparable - albeit advisory - list had been introduced in 1924. Like the German list, the Dutch variety list had also initiated the abandonment of many varieties that were considered to be inferior, as well as varieties that were traded under synonymous names. The Dutch "cleaning" process, however, developed more gradually. By 1935, synonymous denomination no longer occurred (De Haan, 1949:23). From the 353 crop varieties that had been included in the Dutch national variety list in the period 1924-1939, 154 (44%) had been abandoned (De Haan, 1940:12). In 1949, 415 (48%) out of a total of 871 varieties had been eliminated from the Dutch list, including 40 landraces (De Haan, 1949:42).

[10] Pointing to the explosive situation in rural America, the president of the American Farm Bureau Federation O'Neil warned in 1933: "[U]nless something is done for the farmers, we will have a revolution in the countryside in less than 12 months" (O'Neil quoted in Saloutos, 1982:43).

[11] The New Deal policies "... more or less set the pace for [agricultural] policy-making in the post-World War Two decades for Republicans and Democrats alike" (Saloutos, 1982:254).

[12] The support for specific layers of the agricultural community under the AAA was not entirely new. In fact, during the years prior to the New Deal, research, teaching and extension work had been concentrated almost exclusively on the well-managed farm (Saloutos, 1982:261-261).

[13] In order to produce hybrids, a group of parental lines must be formed. This is done by repeated self-pollination of individual plants - a process called inbreeding. Each inbred line is (almost) genetically uniform and different from each of the other inbred lines.

Chapter 4

The Second Agro-Food Order (1930s-1980s) The Emergence of a State-led Crop Development Policy

It was the American long-term strategy for agriculture, developed during the New Deal, that was eventually adopted by the entire Western world. First in Western Europe soon after the Second World War, and later in developing countries, when they had freed themselves from their colonial ties. The strategy involved a state-led transformation of the national rural production sector, which would greatly facilitate the industrialization of agriculture.

This chapter examines the crop development policy that was at the heart of the state-led production strategy for agriculture. How did plant breeding, intellectual property protection for plants, and plant conservation develop during the Second Agro-Food Order? What were the main political controversies that determined their evolution? And how did the crop development policy promote the industrialization of agriculture?

The chapter broadly consists of two parts. Sections 4.2 to 4.4 focus on the crop development policies that were designed, debated and implemented in Western Europe and the USA during the post-war period. The final section (4.5) is devoted to developing countries and the mechanisms that transferred the predominantly American crop development policies to developing countries. We start off, however, with a more general description of the Second Agro-Food Order and the main geo-political tendencies that formed the background of crop development.

4.1 State-led industrialization of agriculture

By the end of the Second World War, little was left of the First Agro-Food Order. The pre-war international trade patterns were completely disrupted by the Second World War and national liberation movements in European colonies in the 1940s and 1950s. In addition, the agricultural crisis of the 1930s had generated a political majority for substituting the defensive, short-term governmental policies on agriculture by a long-term offensive strategy. The American agricultural strategy served as the model.

At the end of the war, the USA emerged as the new economic and military hegemonic power and successfully fostered a world trade pattern free from pre-war protected intra-colonial trade flows. The American position as agro-food producer was strong. The country spearheaded the process of state-supported agro-industrialization, and American private agro-industry was able to expand internationally. Both tendencies, which involved changing trade patterns and agro-industrial internationalization, would largely determine the post-war international division of labour in agriculture (Friedmann and McMichael 1989).

The First Agro-Food Order had been characterized by an international division of labour, in which a colonial trade flow of tropical products and an international flow of temperate products co-existed. The decolonization process of the 1950s and 1960s interfered with both trade flows. Colonial trade in tropical products to Europe and the USA was disrupted, but would proceed as an extension of the inter-*national* trade pattern between independent states. Simultaneously, the de-colonized states became part of the international trade in temperate food products, notably as importers of wheat from the USA. America enjoyed great prestige among the newly independent states, which were receptive to the opportunities of trade and aid the country offered. The U.S. farm sector, for its part, desperately needed new outlets for its agricultural surpluses, when the traditional exports to Europe diminished due to a recovery of European production. The Third World share of world wheat imports grew from 19% in the late 1950s to 66% in the late 1960s (Friedmann, 1992:372).

Meanwhile, an internationally operating agro-industry emerged, initially in the USA, and later also in Western Europe. This industry developed transnational chains of agro-food production. The agricultural sectors of various countries were integrated as a producer of industrial inputs in these transnational production chains. Instead of finished agricultural products, this transnationalization implied the exportation of, for example, maize or soyabean as intermediate products, which are degraded and re-assembled elsewhere into meat or processed food items. Worldwide, farms and farming regions began to specialize in order to become more competitive on the world market.

The Second World War also proved to be a watershed in agricultural regulation. During the First Agro-Food Order, governments in some European countries and the USA slowly started to improve the competitiveness of their national agricultural sector. As explained in the previous chapter, only in Germany and the USA were long-term strategies developed to reorganize and industrialize agriculture. The American New Deal was unique in the sense that for the first time a government became deeply involved in the organization of its national agricultural sector without violating the main principles of the market economy. After the war, it was this American strategy that stood as a model for Western Europe and, to a lesser extent, also for developing countries. This is the reason that, in our opinion, the Second Agro-Food Order can be characterized by the dominance of a *state-led agricultur-*

al production strategy. This production strategy fostered the process of agro-industrialization guided by national political agendas in the Western world, and sustained the new international division of labour in agriculture.

The central role assumed by the U.S. government in reorganizing and industrializing agriculture was part and parcel of an overall economic, interventionist strategy inspired by John Maynard Keynes. Keynesian policy turned the U.S. government into the largest economic actor after 1945. During the post-war period, Keynesian interventionism, including the New Deal agricultural strategy, was 'exported' to Western Europe. With the Nazi regime defeated, and the continent divided into two antagonistic power blocs during the Cold War, Western Europe was amenable to the new American economic principles. Also welcomed was the Marshall Plan. This plan encompassed a massive American economic support programme for Europe that was intended to prevent both a protectionist revival and communist influence in Western Europe. The Marshall aid laid the foundations for an integrated Atlantic economy based on the generalization of the principles developed during the New Deal (Van der Pijl 1984:147-149).

The American agricultural production strategy and its specific crop development policy were introduced into Europe as an element of the Marshall Plan. The new production strategy gained prominence in the 1950s when the agricultural markets of six West European countries were integrated under a Common Agricultural Policy (CAP). The European controversy on the design of the CAP can be considered as a disagreement on the question of how to implement the new agricultural production strategy. The Commission of the European Economic Community (EEC) strongly favoured a European version of the agricultural New Deal. The Conference of Stresa in 1958 (at which the CAP was founded) as well as the Mansholt Reports of 1961 and 1968, reveal the Commission's objectives: higher agricultural productivity, more governmental control over agricultural output, lower internal agricultural prices, and the removal of 'non-viable' farms (*cf.* Tracy, 1982:227-258; Neville-Rolfe, 1984:185-245).

The state-led agricultural production strategy succeeded in raising agricultural productivity, both in the USA and Europe, due to the specific crop development policy. This policy consisted of three main elements: (a) massive public investments in agricultural research[1] (Alston *et al.*, 1998:55); (b) the creation of hybrid varieties and the adoption of plant breeders' rights legislation; (c) the active adjustment of conservation to plant breeding objectives. Particularly the latter two elements are explained in detail in sections 4.2 to 4.4.

The new crop development policy paid off well. The yields of maize, wheat and other crops more or less doubled in the USA between 1940 and 1970 (Figure 4.1). The improved genetic constitution of crops can be an important factor in yield increase. It has been claimed that 70 to 80% of the total maize yield gains between 1930 and 1980 can be ascribed to better varieties (*cf.* Duvick, 1977:187; Duvick, 1984a:15). Better varieties increased wheat yields by 17% between 1958 and 1980 (Schmidt, 1984:89).[2]

Figure 4.1 Yield increase of maize, wheat, and all crops between 1910 and 1970 in the USA (all crops 1967 = 100)

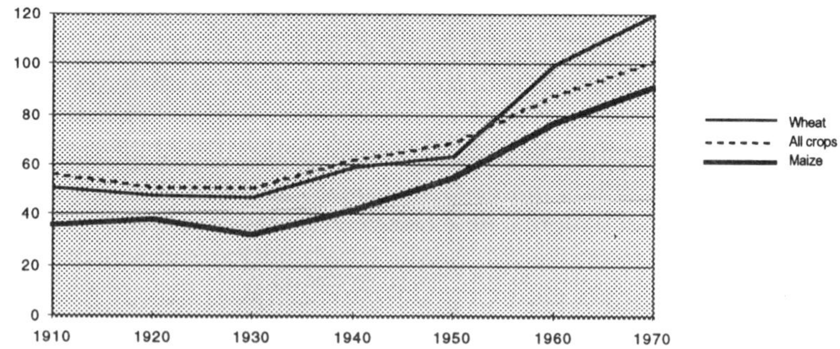

(Data derived from McMullen, 1987:24)

The state-led industrialization of agriculture in the USA and Western Europe involved a drastic reorganization of the farm sector and aroused fierce opposition. Due to the political power of the farming lobby, governments on both sides of the Atlantic were confronted with social costs incurred by industrialization. Both the U.S. government and the EEC Commission were forced to keep internal prices for the main crops well above the world market level. The price policy delayed the removal of smaller 'non-viable' farms. Although farm employment diminished rapidly, with superfluous farm labour being absorbed by the expanding industrial and service sectors[3], the rural exodus did not reach the extent envisaged. The combination of drastic productivity increases and relatively high national prices, resulted in the accumulation of food surpluses which facilitated the export position of the American and European agro-industry.

At the same time the American crop development policy was adopted in Europe, it was also introduced into a number of developing countries through the finance and expertise offered by the U.S. government and American philanthropic foundations. The 'export' of the American crop development policy would initiate a "Green Revolution" in many developing countries. It was intended to prevent a red revolution by increasing food security, but also offered the U.S. government the opportunity to establish a centrally organized global conservation system.

The American crop development policy proved to be rather successful in several developing countries. The expansion of agricultural production enabled many countries to improve national food security and take part in the international division of labour in agriculture. Apart from importing non-traditional food, developing countries started to export agricultural products, basically as participants of the transnational chains of production. In respect of global plant conservation, the Green Revolution has also been successful, as we point out in greater detail in section 4.5. The success of the crop development policy in developing countries had a reverse side, however, because of its considerable social and environmental consequences. Chapter 6 is entirely devoted to the opposition that Green Revolution agriculture provoked in two Latin American countries.

In sum, the Second Agro-Food Order had two characteristics. On the one hand, it consisted of a specific international division of labour in agriculture dominated by a transnational agro-industry. On the other hand, the division of labour was accompanied by a specific agricultural production strategy that was developed on both sides of the Atlantic Ocean. The strategy was directed towards an industrialization of agriculture under guidance of the national government. It entailed a specific crop development policy aiming at an increase in productivity of bulk crops, an increase of private investment in plant breeding, and the avoidance of large social costs.

4.2. Diffusion of the inbred-hybrid breeding method

As explained in the previous chapter, the U.S. government firmly encouraged the inbred-hybrid method as the main breeding policy in maize. Because of the extensive research investments, hybrid varieties eventually began to out-yield the open-pollinated varieties and thus became commercially attractive to American farmers. In 1930, a few years after the first maize hybrids had been introduced, 0.1% of the American maize acreage was planted to hybrid varieties. By 1945, this figure had risen to 53%, and by 1960 to 94% (Kloppenburg, 1988a:120). Hybrid maize was crucial to the development of the private seed industry in the USA. Hybrids effectively eliminated the double use value of plant varieties (the use both as a crop and as propagating material), and gave private breeding firms the opportunity to obtain higher returns on their investment. Prior to the arrival of hybrid maize in the 1930s, there were around 40 seed companies operating in the USA. Around 1980, this figure had risen to around 340 companies, most of them dealing with hybrid maize seed (Doyle, 1985:35).

The inbred-hybrid method was also applied to sorghum, a popular new feed grain after the war. Hybrid sorghum boomed in the late 1950s. By 1960, 70% of all sorghum acreage in the USA was planted to hybrids. Hybrids were later also developed in other field crops, such as sunflower and sorghum, but also in pigs, cattle, and chickens. The first hybrid chickens were marketed by Pioneer Hi-Bred in 1942 (Doyle, 1985:43-44).

Research for the development of hybrid wheat has also been under way since the 1940s. Several seed companies, such as Cargill, Northrup King, DeKalb and Pioneer Hi-Bred invested millions of dollars on hybrid wheat research in the early 1960s. The results were poor, however. The genetic and physical make-up of the wheat plant makes it less suitable for hybridization. Moreover, the barriers for wheat hybrids to enter the market were raised considerably in the second half of the 1960s, when new, open-pollinating wheat varieties were released that were high-yielding. This development, and the prospect of legal plant protection (see section 4.3.4) induced a major reduction in hybrid wheat research programmes at the end of the 1960s (Knudson and Ruttan, 1988).

In Europe, the phenomenon of hybridization had been known since the early 1930s and American hybrid maize seed was imported regularly (Van der Have BV, 1979:62). The interest in hybridization was modest, however, since the main carrier of that technology, maize, was of little importance in Europe. Ever since Columbus carried maize kernels from Cuba to Spain, maize has been grown in the Mediterranean region, but only on a small scale. Climatic circumstances prevented maize production in Northern Europe. The traditional grain and forage crops in this region are oats, rye, barley and wheat. Only when early maturing lines that were adjusted to the climate in Northern Europe became available, did interest in maize hybrids increase.

The U.S. seed industry was eager to export hybrid maize seed to war-torn Europe. The UN Relief and Rehabilitation Administration (UNRRA) and later the FAO facilitated the introduction of American hybrid maize varieties (FAO, 1959a:3). Willing to adopt the new technology, European seedsmen crossed the Atlantic and visited American maize companies and institutes in the late 1940s (Van der Have BV, 1979:62). Inbred lines were offered by the U.S. research stations within the framework of the Marshall Plan (CEBECO, 1963:95). In 1953, six% of the European maize acreage was planted to hybrids (Kjaer, 1955:86). By 1959 the use of hybrids had risen generally, up to 98 and 99% respectively in Belgium and The Netherlands (FAO, 1959a:3)

Most successful in northern Europe was silage maize (or 'green corn') which is harvested before the plant reaches maturity. The whole plant is harvested, dried, ensiled and used as fodder. Silage maize proved to be excellent as feed, as its high protein content allows for an efficient transformation into meat. European governments stimulated national maize production with the objective of reducing the import of American fodder maize. The European market of silage maize began to expand in the late 1960s, at the expense of the markets of traditional local feed crops (Van der Have BV, 1979:63).

Nevertheless, since maize was only a minor crop in Europe, the effect of maize hybrids on the private seed firms in Europe was less significant than in the USA. Even though the inbred-hybrid method was also applied to European field crops, particularly sugar beet, governments in Europe supported the private seed industry in a different way. In Europe, the double-use value of plant varieties was restricted by legislation rather than by technology.

4.3 International recognition of plant breeders' rights

Hybridization was a technique that offered breeders protection against unauthorized propagation of plant varieties. Hybrid varieties, however, could only be produced in a limited number of field and vegetable crops. For the majority of field crops and ornamental plants breeders sought legal solutions to the double-use value of plants. Particularly in Europe, the development of Plant Breeders' Rights (PBR) was a core element of the crop development policy during the Second Agro-Food Order.

PBR protection evolved in Europe and was in fact the outcome of a conflict of interests between private plant breeding firms, industrial patent departments, and agricultural governmental authorities. The plant breeders aimed at patent coverage for their creations, while industrial patent lawyers and governmental patent offices showed little interest in the issue and were concerned that patent protection of plants would weaken the entire patent system. The agricultural governmental authorities, for their part, intended to facilitate private investment in plant breeding, but disliked the idea of legal monopolies in that sector.

The European controversy on the question of how the unauthorized propagation of plant varieties could best be controlled started in the 1920s and would only be resolved in 1961, with the adoption of the Convention for the Protection of New Varieties of Plants (UPOV). This section examines the route to UPOV: the first plant breeders' protection systems in European countries, the political conflicts between industry and plant breeders, the adoption of UPOV and, finally, the introduction of PBR in the USA.

4.3.1 European initiatives in protecting plant varieties

The way private breeders used to protect their plant varieties against unauthorized propagation was essentially similar throughout Europe. Since the 1920s, their protection relied on two regulatory mechanisms: the plant variety catalogue, and seed certification. National plant variety catalogues were established in several countries with the aim of reducing the 'varietal jumble'. Although the requirements and function of the catalogue varied among countries, registration of both the name of the new variety and its breeder was common.[4] Seed certification was introduced to control the quality of the seed sold on the market, but it was also used by the authorities to check whether the varietal name printed on the label matched the contents of the seed bag. Since no other breeder could register a variety under an existing name, and since only seeds of registered varieties were allowed on the market, the combination of measures offered the breeders some protection. It was, however, the varietal name rather than the variety itself that was protected.

In countries where plant breeding was dominated by the public sector, such as Sweden, Great Britain, Ireland and Switzerland, additional opportunities restricting unauthorized propagation of plant varieties did not become an element of governmental crop development policy until the 1960s. Only in those countries where plant breeding was mainly in private hands, such as France, Germany and The Netherlands, did the state offer supplementary protection in the 1930s, even though it was contested within the agricultural community. Breeders' organizations endeavoured to expand the control of the propagation of their varieties, while organizations of farmers and of seed producers opposed any limitation of their seed saving and producing opportunities.

This ongoing tug of war resulted in the French seedsmen protecting their varietal name or logo under trademark law (Strauch, 1972:33). German breeders were able to patent their plant varieties in the mid-1930s until the Nazi regime interrupted the plant patenting route and offered breeding firms *de facto* protection of their varieties by the authoritarian control of the seed market. In 1953 the West German government adopted the Law on the Protection of Varieties and Seeds of Cultivated Plants, which offered breeders the exclusive right to produce and market the seed of their varieties (Weibull, 1955:120).

A similar, although weaker, plant variety protection system had been introduced in the Netherlands in 1937. The system, jointly created by breeders' and farmers' organizations, offered breeders exclusive rights on the marketing of seed of their own varieties, except for main food crops. For these, exclusive marketing rights were only allowed on the first generations, known as elite or original seed. Although the Dutch system offered breeders less control on the propagation of their own varieties than the German legislations, from 1934 and 1953, anyone who propagated the variety for commercial purposes was charged a levy for every hectare of seed that was certified. The revenues were allocated to a Breeders' Fund from which breeders were remunerated. The more seed of their varieties that was certified, the bigger the share of individual breeders in the Fund (Addens, 1952:174). Some years after its introduction in 1937, the Dutch breeders' remuneration system received government support through the adoption of the 'Breeder's Decree of 1941' (Board for the Plant Breeder's Right, 1946). [5]

The interest in legal protection for plant varieties did not remain confined to national territory. The small national production base and market had induced the seed industry since its inception to sell its material throughout Europe. Plant breeders realized, therefore, that legal protection of breeders' creations required international recognition in order to be effective.

One of the first times that the issue of international plant variety protection was raised was probably during the 5th International Congress for Hereditary Science, in 1927 in Berlin. At the same Congress where Vavilov presented his theory on the World Centres of Origin of Cultivated Crops, a commission was established to prepare a proposal on plant variety protection (Van Harreveld, 1934:28). In the subsequent decade, the European agricultural establishment showed its sympathy for property rights for breeders and for an international agreement in this field, on various occasions. The issue was discussed during the 1931 congress of the *Association Internationale des Sélectionneurs de Plantes*, the congresses of the Rome-based International Agricultural Institute in 1936 and 1940, and the International Agricultural Congresses of 1931, 1934, and 1937 (Verhulst, 1947:36-41). During the 1937 congress, a resolution was adopted in which the International Agricultural Institute was requested to organize a meeting of experts which would prepare an international convention on the protection of breeders' creations (Addens, 1952:184). None of the resolutions and proposals adopted during the agricultural meetings in the 1930s gave any practical results, however.

4.3.2 Industry rejects patent aspirations of European plant breeders

The establishment of a special lobby organization, the International Association of Plant Breeders for the Protection of Plant Varieties (ASSINSEL), illustrates the strong interest of the European breeding firms in an intra-European plant variety protection system. ASSINSEL was founded in 1938 in Amsterdam by seedsmen from France, the Netherlands, Denmark and Belgium. The organization was soon joined by private seed industry representatives from other European countries.[6] Apart from a disruption during the Second World War, ASSINSEL worked continuously on the establishment of a European system for the protection of intellectual property rights for breeders. After the war, the need for protection of plant varieties was intensified by new internationalization strategies of the European seed industry. Many seed firms of North-West European countries started to propagate seed of their own varieties in Southern European (and also African) countries, because of the sunny and dry weather conditions during flowering time and the availability of cheap labour (Thielebein, 1955). It was not yet clear for ASSINSEL members, however, which type of protection was more appropriate. During the annual conferences until the mid-1950s, two options were discussed: trademarks and patents.

Trademark law had generally been the system to which many private breeders resorted for the protection of the name of their new varieties. The French seedsmen in particular relied on a combination of trademark law and compulsory registration in the national variety list. The intention of ASSINSEL's president Ernest Tourneur was to use the French system as the international standard, but this turned out to be impossible. The obligation to register new varieties in the national catalogue only existed in a few countries, while there was still no consensus on the synonyms of plant variety names in the different European languages (ASSINSEL, 1947:5,9). Tourneur eventually dropped the trademark option in 1950 in favour of patents (ASSINSEL, 1950:4).

Patents could offer a more secure protection and were - at least with respect to industrial inventions - already internationally recognized through the Paris Convention. Discussions at ASSINSEL's annual congresses show, however, that the breeders realized the limited value of patent law for plant breeding. Only plants whose progeny were identical to the parents would have been eligible for protection. Because characteristics of cross-pollinating plants change after each reproduction cycle, it was not possible to provide an exact description of these plants - a limitation that precluded patent protection. Hence, alternative means of protection had to be sought (ASSINSEL, 1949a). But at that time, a specific breeders' rights system was not yet considered as an option. The German PBR legislation was only enacted some years later,[7] while the association was presumably not impressed by the Dutch system.[8]

By opting for patents, ASSINSEL entered into a controversial area. On the one hand, ornamental plants, mainly roses and carnations, had been considered to be patentable in Germany since the 1930s, in France since 1949, in Italy since 1951, and in Belgium since 1958. In Italy and France the establishment of a Plant Patent

Act was even considered, like the one enacted in the USA in 1930 (Heitz, 1991:29-30). On the other hand, the Agricultural Ministries in Europe generally rejected plant patents. In Switzerland, the Netherlands, Denmark and Great Britain, patents for plants were deliberately not issued. The main opposition to the patent aspirations of the plant breeders during the Second Agro-Food Order in Europe, however, came not from Ministries of Agriculture, but from the industry.

The private crop development industry in Europe generally consisted of small, rural, family-owned firms. Until the 1970s, involvement of large industrial capital in this sector did not occur, because the technical control of the hereditary mechanism of plants was still limited. This factor dimmed the prospects on returns, but also prevented the possibility of 'invented' plants meeting the requirements for patent protection. The mechanical and chemical industries, which were the main users of the patent system, had therefore little interest in the efforts of plant breeders.

The first time that industry's patent lawyers showed their reluctance to accept plant patenting was during the 1932 congress of their professional organization, the International Association for the Protection of Industrial Property (AIPPI). AIPPI had been established at the end of the 19th century to promote patent protection and incorporated patent lawyers of all major industrial enterprises and national patent offices. The 1932 AIPPI congress disapproved of a proposal for plant patenting submitted by a German lawyer, because it was thought that the patenting of plants would endanger the entire patent system (AIPPI, 1933:170).

The AIPPI position on plant patents remained rather static throughout the 1940s and 1950s. Although differences of opinion existed among the national groups,[9] the industrial association generally did not intend to exclude inventions in the field of plant breeding from patent protection. AIPPI was concerned that a weakening of the patent requirements, which would have been necessary in relation to plants, would negatively affect the patent system as a whole. It argued that the basic criteria for patent protection - that there is an invention and that this invention can be reproduced by others - could not be met in the case of plant varieties. For this reason the overwhelming majority of AIPPI members voted against the protection of plants by patents in 1954 (AIPPI, 1954:63; ASSINSEL, 1954:5). The industry's lawyers were of the opinion that plant breeders deserved protection for their creations, but through a legal system different from the patent system. It proposed to protect plant varieties by a system specifically designed for that purpose. At an earlier stage AIPPI had encouraged ASSINSEL to take the initiative in this respect (ASSINSEL, 1947:5).

The industry's negative position on the patenting of plants was maintained until the 1970s, as is reflected in the outcome of several international patent law agreements. During the 1958 conference on the revision of the Paris Convention in Lisbon, the experts (mainly patent office directors), refused to discuss the issue of plant patenting (Heitz, 1991:34). Neither was the question resolved with the convention on the unification of patent laws of European countries, adopted in 1963 in Strasbourg. The Strasbourg Convention left the decision on granting of patents for plants to the member states.[10] Ten years later, in 1973, when the European Patent Convention (EPC) came into force, the patenting of plant varieties was explicitly rejected.[11] The European policy not to offer patent coverage for plant varieties was facilitated by the emergence of a new legal intellectual property system, specifically designed for the protection of plant varieties, commonly known as UPOV.

4.3.3 The foundation of UPOV

While the industry was pushing plant protection away from the area of patenting, the agricultural departments of European governments attempted to establish a new protection system under their control. The agricultural authorities seized the opportunity in 1956, when the French government organized an international conference on the issue of plant variety protection. The French conference was in fact a joint initiative of ASSINSEL and the French Department of Agriculture, which had been closely collaborating since the late 1940s. Most participants at the conference also originated from the agricultural sector.[12]

Four years of negotiation resulted in the International Union for the Protection of New Varieties of Plants (UPOV), signed in 1961. The UPOV Convention provides minimum standards which member states have to meet in their national legislation. The important implication of UPOV was that in Europe the creations of plant breeders were not protected by patents, as ASSINSEL desired, but by a legal system exclusively designed (i.e. *sui generis*) for the protection of the property rights of plant breeders, commonly referred to as plant breeders' rights (PBR). The plant patent advocates were not left with empty hands, however. UPOV member states were still offered the opportunity to protect a plant variety by a patent, but double protection of plant varieties was prohibited. Member states could provide either patent or PBR protection for one and the same botanical genus or species.

The foundation of UPOV made it much easier for European plant breeders to obtain legal protection, because the requirements are less demanding for PBR protection than for patent protection. Patent law requires an *invention*, while under PBR law any *new plant variety* is in principle eligible for protection. Patent law also requires that the invention is reproducible by a skilled person, while the criterion for PBR protection is that the varietal characteristics are defined and described. An invention must be non-obvious and novel worldwide. Since most varieties are obvious adaptations of existing varieties, PBR law requires only that the important varietal characteristics are clearly distinguishable from other varieties that are commonly known. Finally, PBR law requires that the variety is sufficiently genetically uniform, while the variety's essential characteristics must remain stable, i.e. must remain true to its description, after repeated reproduction.

However, these lower protection requirements came at a price, in that the scope of PBR protection is narrower than that of a patent. Whereas patents protect the inventor against *all* unauthorized commercial use of the invention, a PBR certificate entitles the breeder only to prevent unauthorized commercial *propagation* of plant varieties.

The narrow scope of PBR protection had two significant consequences. In the first place, PBR did not prohibit a third person from using the protected variety to breed a new one, selling this new variety and protecting it under PBR law. From a legal point of view, new varieties remained independent from their source varieties. The resulting opportunity for breeders to freely use protected varieties for breeding has become known as the 'breeders' exemption'. Secondly, the scope of PBR did not extend to the propagation of protected varieties for replant-ing purposes by farmers. This 'on-farm seed saving' was considered not to be a commercial activity, thus falling beyond the scope of PBR. Later, on-farm seed saving would be referred to as the 'farmers' privilege' (or the 'crop exemption').

Both the terms 'farmers' privilege' and 'breeders' exemption' suggest that they are derogations from the breeders' right. The absence of these terms in the original UPOV Convention, however, indicates that the European Departments of Agriculture, which prepared the convention, did not want to extend the scope of the breeders' right to on-farm seed saving and the use of a variety for further breeding. The fundamental objective of UPOV was only to *limit* one of the triple-use values of plant varieties. PBR-protected varieties could be used for growing a crop, for creating new plant varieties, for on-farm propagation, but not for unauthorized commercial propagation.

The UPOV convention was signed by several European countries in 1962, but only came into force 6 years later, in 1968, when three countries ratified the convention: the United Kingdom, West Germany and The Netherlands. It would take several more years before the system was operational in other European countries, a delay partly due to an ongoing patent versus PBR controversy at national level. France, for example, ratified the convention only in 1970, almost 10 years after signing it. Sweden acceded to UPOV in 1971, Belgium in 1976, Italy and Switzerland in 1977.

4.3.4 Adoption of breeders' rights in the USA

Whereas the USA was not involved in the UPOV process, it did not lag behind and adopted the Plant Variety Protection Act (PVPA) in 1970. The fact that plant variety protection was available in the USA earlier than in most European countries is quite remarkable. At the time UPOV was established, the American seed sector still opposed legal protection for sexually-reproducing plants (Laclavière, 1972:54). Seed companies dealing with hybrid varieties had little to gain from PBR. Moreover, American breeders were reluctant to accept the European PBR system because it was rooted in governmental seed trade regulation, involving compulsory registration in the national variety list and seed certification. They were concerned that the UPOV system would give the government the authority to decide which variety was allowed onto the market and which not (Fowler, 1994:104).[13] Nevertheless, in 1963 the American Seed Trade Association (ASTA) established a committee to assess plant variety protection (PVP).[14]

Advocates of PVP were relative newcomers in the area of plant breeding: the seedsmen involved in the breeding of forage crops, who desperately required income to finance their research programmes (White, 1964:70-71). Moreover, the expectations of hybridization had not materialized. In several crops, hybrid varieties were developed, but not in major commercial crops such as wheat and soyabean (Bugos and Kevles, 1991:14).

The question of legal protection for plants became topical in 1966 when a revision of the entire patent law, including the PPA, was in preparation. The President's Commission on Patent Reform published its report, and recommended that the 1930 Plant Patent Act be repealed. The Commission considered the patent system not to be the proper vehicle for the protection of plants and seeds, regardless of whether they were sexually or asexually reproducing. Nevertheless, the Commission did acknowledge the valuable contribution of plant breeders and urged further study to determine the most appropriate means of legal protection for plant varieties (Doyle, 1985:56-57).

When in 1968 the Senate Judiciary Committee embarked on a series of hearings on the Patent Reform report, ASTA seized the opportunity not only to oppose the patent reform recommendation, but also to push for an extension of the coverage of the PPA to sexually reproducing plants. Despite the support of powerful organizations, such as the National Cotton Council and the National Canners Association, ASTA's amendment failed. The USDA was "particularly firm in opposing plant patents" and argued that asexually propagating varieties would not remain true to their description after repeated reproduction (Doyle, 1985:58). The American Farm Bureau Federation also advised negatively on the issue, arguing that the interests of breeders of sexually-reproducing plants were sufficiently served by seed certification programmes (Bugos and Kevles, 1991:16).

When the option of general plant patenting under the *aegis* of the Patent Office had failed, ASTA was prepared to accept a breeders' rights system for sexually reproducing plants administered by the USDA, in addition to the PPA. The legislative course in Congress changed accordingly, from the Judiciary Committee to the Agriculture Committee. A PVP bill was drafted by representatives of the public and private seed sector and the USDA, and did not meet significant opposition on the part of farmers, consumers and Congress. The bill passed Congress in 1969 and was signed into law in 1970 (Doyle, 1985:60-66).[15]

Summary

Throughout the Second Agro-Food Order, American and European plant breeding firms tried to restrict the dual-use value of plants by means of technology and legislation. Hybrid varieties were developed in maize and in some other crops. But hybridization did not offer a solution to unauthorized propagation of plant varieties of major commercial crops, such as wheat, soyabean, flowers and fruit. Attempts to patent new varieties were thwarted by the industry. Until the 1970s, the major industrial sectors had little interest in plant breeding, which was basically a rural-based business in the hands of family-owned firms. The industrial position left plant breeders no other choice than to accept the plant breeders' rights system.

4.4 American versus European conservation strategies

The large-scale conservation of landraces as breeding material is the third element of the state-led crop development strategy during the Second Agro-Food Order. Like the diffusion of hybrid varieties and the emergence of breeders' right protection, considerable differences existed in the strategy for landrace conservation between the USA and Western Europe.

On both sides of the Atlantic, the conviction grew that landraces constituted an important resource of genetic information for plant breeding and agriculture. As we show in section 4.4.1, it was only in the USA that the state took a leading role in adjusting conservation strategies to plant breeding policies. The growing involvement of the U.S. government in conservation in the 1950s led to a far more centralized conservation strategy than in Europe. This allowed the U.S. government to adjust conservation to long-term, national plant breeding policies.

In Europe, the conservation of landraces would largely remain in the hands of breeders themselves. This tradition, in combination with diverging national considerations, prevented the emergence of a common, centralized European conservation strategy (section 4.4.2). Long-term national conservation strategies emerged only in the 1980s and 1990s, while a common European conservation strategy is still under construction. This explains why in Europe an adjustment of conservation strategies to long-term plant breeding policies never emerged as it did in the USA.

4.4.1 Central seedbank system for American 'new crops research'

Since the USA is poor in native genetic resources, the uninterrupted supply of exotic plant material has always formed a key element of national crop development. From its establishment in 1862, the USDA systematically imported landraces and wild relatives of the most important crop plants (*cf.* Wilkes, 1983; Yeatman *et al.*, 1984; Cox *et al.*, 1988). In the first two decades of the 20th century, a national network of Federal Plant Introduction Stations was established (see section 2.2.3). In order to ensure ready access to germplasm in the future, the U.S. government further extended this national network after the Second World War. The juridical basis for this extension was provided by the 1946 Research and Marketing Act, which included an amendment to the Bankhead-Jones Act of 1935. The 1946 act resulted in the construction of four Regional Introduction Stations and a National Potato Introduction Station. All were built between 1947 and 1953. Some years later, in 1957, the national 'mega' seedbank emerged: the National Seed Storage Laboratory (NSSL) in Colorado (Brown, 1984:32). The NSSL is still generally known as the world's largest *ex situ* cold storage facility.

The extension of the seedbank network was inspired by a broader and long-term view on the importance of landrace conservation for agriculture. Apart from serving as a genetic backup to augment the overall performance of crops (in terms of yield, resistance, etc.), the landrace collections were intended to adapt plant characteristics to the specific requirements of the agro-industry. The 1946 act repeatedly speaks of the conservation of plants for "agricultural or industrial use". Reference is made to the food processing industry, as well as to the "chemical or manufacturing" industry (Wilkes, 1983:148). The screening of landraces for traits with a variety of (potential) agro-industrial uses expanded the scope of conservation and collection.

The extensive American seedbank network also played a valuable role in solving a contradiction in the agricultural production strategy. The advances in crop development had greatly contributed to the rise in agricultural productivity. The strong lobby of farmers and agro-industry, however, prevented overall production from being adjusted to national demand. Through the Marshall plan, American excess production could be exported to Europe in the aftermath of the Second World War, but after the Korean War had ended, American food surpluses accumulated rapidly (Cochrane and Ryan, 1976:14-15).

It is against this background that, in 1959, the American Association for the Advancement of Science organized a symposium on Germ Plasm Resources. Two USDA researchers, Quentin Jones and Ivan Wolff, presented a paper titled: Using Germ Plasm for New Products, in which they proposed a "new crops research" policy for public agricultural science (Jones and Wolff, 1961:267). The policy had two objectives. The first was the introduction of new crops to "eliminate expected future increments of surplus of wheat". Particularly soyabean, at that time a relatively new crop, was recommended as a candidate. The other objective was to find industrial rather than food markets for agricultural crops.

> "It ... appears evident that we should concentrate on those industrial needs for which plant products have an inherent competitive advantage over materials of non-agricultural origin. This means that our efforts should be centered on those plant constituents, such as seed oils, proteins, gums and fibrous cellulose which ... can be adapted to special uses." (Jones and Wolff, 1961:267).

The growing possibilities of using "germplasm"[16] for new agro-industrial needs required a new conservation strategy. In the 1960s, the existing national seedbank network was expanded again with twelve new seedbanks. The extended network, in turn, became part of the National Plant Germplasm System (NPGS). Besides seedbanks, the NPGS encompassed a network of research units and agencies, which jointly introduced, evaluated and distributed seed and plant material (Hougas, 1984:17-20). Quentin Jones, one of the architects of the new crops research strategy, was appointed as the NPGS director.

The NPGS expresses an important characteristic of American crop development policy during the Second Agro-Food Order: i.e. a strong synergy between plant breeding and conservation. In the post-war period both branches became locked within one bureaucratic system under direction of the Agricultural Research Service and the USDA. The NPGS facilitated an adjustment of conservation strategies in accordance to the changing demands of plant breeders. To this end, an advisory National Plant Germplasm Committee was established, which consisted of representatives of the private sector (the Council of Commercial Plant Breeders) as well as the public sector (the Cooperative State Research Service) (Hougas, 1984:21).

4.4.2 Fragmented European seed collections

During the Second Agro-Food Order, European conservation policies contrasted quite sharply with those of the USA. While American crop development policy was directed by the USDA , European conservation efforts were fragmented because of the absence of an overall European policy. After the war, Western European countries followed their own, national conservation policy, while many collections remained in the hands of private breeding firms serving *ad hoc* needs.

Until the 1940s, conservation in Europe had merely been an activity of botanists who collected for the benefit of public botanical gardens. The collections were used for academic taxonomic study, but their most important function was as a 'transfer point' for botanical exchanges within the colonial empires. The interest in central seed collections of temperate crops has never been as extensive as in the USA. In part, this can be explained by the prevailing interest in tropical, colonial agriculture. Other reasons, however, lie in the long-standing history of plant introduction in Europe, which goes back several millenniums, and the fact that the world's main centre of origin of grain crops, the Mediterranean basin, was at hand. In this sense the very active and central conservation policy of settler states, notably the USA, can be considered as an attempt to catch up with agricultural history.

The importance of collecting landraces for breeding was well known among European plant breeders since Vavilov had presented his findings.[17] A European central conservation policy, however, would prove difficult to materialize. National considerations, and the existence of a relatively strong private breeding sector, for a long time prevented something like a 'common conservation policy' form developing in Europe. Various initiatives in this direction failed.

An early proposal for the establishment of an "inter-European system for plant exploration, collection and introduction" was submitted in 1958 but never materialized. None of the countries that were initially interested (Denmark, France, Germany and Norway) eventually wanted to host more collections than they already had. The seedbank curators involved in the negotiations refused to have their collections swamped with material that was beyond the needs of the national breeding programmes. The FAO also had to turn down the request to host the collection, because European donor countries were not prepared to pay additional funding to this UN organization (FAO, 1958a:3; Whyte, 1959). Another initiative to set up a European conservation system came from the European Association for Research on Plant Breeding (EUCARPIA) in 1966. This too was bound to fail. In 1962, EUCARPIA began to lobby (also within the FAO) for a permanent bureau for exploration and introduction, and the establishment of "various germplasm stations in various parts of the world" (Hawkes and Lamberts, 1977:1). One year later, EUCARPIA proposed to establish a cooperative germplasm conservation project between Germany, the Netherlands, and the UK. The proposal failed because thc UK government was not interested, while support from the Organization for Economic Cooperation and Development (OECD) was refused for financial reasons. Eventually, the German and Dutch governments agreed to establish a joint storage facility for seed potatoes and planting material in Braunschweig (Hawkes and Lamberts, 1977:1).

Instead of creating a centralized conservation network, national governments in Europe left conservation to those who directly used seed: the breeders. In fact, the maintenance of 'working collections' of germplasm by plant breeders of private firms and of public institutes served as an alternative, decentralized conservation 'strategy'. Working collections met the short-term and direct germplasm requirements of national private and public breeders, and were generally located close to the breeder. The variability of conservation practices among Western European countries is illustrated in the following brief discussion of the post-war national conservation initiatives of France, Great Britain, the Netherlands, and West Germany.

In *France*, the public working collections were spread over three agricultural research institutions: the National Institute for Agricultural Research (INRA), the Centre for International Collaboration in Agriculture (CIRAD), and the French Institute of Scientific Research for Development and Cooperation (ORSTOM). The plant breeders in these institutions kept the responsibility their 'own' working collections, which were passed on to succeeding generations of plant breeders (Busch, 1995:134). This system apparently functioned well. A hidden strength of the French conservation system was the landraces on which many small farmers still relied until the 1960s. By maintaining personal contact with the farmers in the region, French breeders had ready access to these landraces. It would take INRA until the 1990s to design a centralized conservation policy (personal communication Ph. Auriau, INRA, 1997).

As discussed in chapter two, the colonial Kew Botanical Gardens remained for a long time *Great Britain's* most important collection of seed. Kew Gardens, however, was not designed to carry out breeding research for temperate crops in the home country, but served as a transfer point in the colonial botanical exchange system. Collections of temperate crops were established in the 1950s, such as the Commonwealth Potato Collection at Dundee and the Barley Collection at Cambridge, but they mainly served academic and taxonomic purposes, rather than those of plant breeders (personal communication J. Hawkes, University of Birmingham, 1997). Other, more 'informal' collections were held under the responsibility of the Ministry of Agriculture and the Agricultural and Forestry Research Council. The remainder of the British collections consists of a plethora of small public and private working collections. Until the late 1980s a national conservation policy was absent (Hardon, 1988:9).

The picture of fragmented conservation initiatives also applies to *The Netherlands* where the government deliberately confined its role in plant breeding to support private firms. Some of the Dutch working collections, especially those containing potato breeding material, were quite large and international in scope. The lack of a centralized conservation policy caused some collections to be simply thrown away when they were no longer useful for the private breeding programmes. The abandonment of the entire collection of Dutch spelt landraces in 1951 (Zeven, 1996:339) may serve as an example. Other collections, among them considerable parts of the field collections of the Wageningen University, were lost in 1940, as they happened to lie in the middle of the Dutch-German front line. The Dutch Centre for Genetic Resources (CGN) set up a centralized seed bank policy in 1984 (Van Soest and Van Hintum, 1995:7-10).

The exception to the fragmented conservation infrastructure in Western Europe were the *West German* collections. Before the Second World War, considerable plant collections had been established by the Kaiser Wilhelm Institute (KWI) in view of Germany's strategy towards food autarky, but the war had taken its toll on them. Some collections were destroyed, some fell into British hands, and some were (re-)captured by the Soviets (later forming the basis of the East German seed bank in Gatersleben). Only a part of the original KWI collection could be maintained in West Germany. The remaining collection was of little interest for German plant breeders because the character of the German agriculture changed rapidly after the war. Due to the division of the German empire, West Germany lost its traditional grain and sugar beet production area to East Germany, while livestock production gained importance (Andrews *et al.*, 1979:12-13).

Flitner (1995) remarks that post-war German conservation activities were strongly linked to those of the USA. The leader of the U.S. Point-Four[18] Programme, Waldee, personally took an interest in the conservation work of the KWI. The Point-Four programme financed a large, joint USDA/KWI expedition in 1952-53 for the collection of wheat samples in Iran (Flitner, 1995:125-139). Another KWI expedition to the Andes in 1959 was financed by the Rockefeller Foundation, the German Scientific Society (*Deutsche Forschungsgemeinschaft*) and several large German pharmaceutical and chemical companies (IG Farben, Leverkusen, and La Roche) (Flitner, 1995:140-148).

If a centralized European conservation policy existed during the Second Agro-Food Order, it was behind the Iron Curtain. After 1964, *East European* countries integrated into the Council for Mutual Economic Aid (COMECON) established a large seed exchange system coordinated by the VIR in Leningrad. Although the USSR could not resume its key role in international conservation, due to the outbreak of the Cold War, West European and American geneticists and conservation specialists continued to visit East European and Soviet countries in the 1950s and 1960s, albeit less frequently (Loskutov, forthcoming). Integrated East-West conservation strategies were intensified again only after 1989.

Summary

The traditional American dependence on the import of landraces of temperate crops culminated in a strong state-led conservation policy after the Second World War. The approach of the USDA gradually evolved into the National Plant Germplasm System, which maintains the most advanced conservation facilities in the world. This system allowed the USDA to adjust conservation to new plant breeding objectives. The European situation was entirely different. The larger involvement of the private sector in crop development, as well as the ready availability of temperate landraces in the field, resulted in a far more fragmented conservation 'strategy'. Until the 1980s when governments individually set up national conservation strategies, the bulk of the European collections was still maintained in individual working collections of private and public plant breeders. As will be shown in the next section, the centralized U.S. conservation system was relatively easy to replicate in developing countries.

4.5 American crop development policy transferred to developing countries

In the second chapter of this book we examine why and how, during the First Agro-Food Order, governments in Europe and the USA began to design and implement crop development policies. To maintain national agricultural competitiveness, governmental authorities started to regulate the seed market, control the quality of the seed, and fund agricultural research institutes. During the Second Agro-Food Order the involvement of the government in crop development increased considerably. Public agricultural research expenditure was increased in the USA, and most West European countries followed the American example. A PVP system was also established on both sides of the Atlantic, but only in the USA did the government play a central role in plant conservation.

Shortly after the Second World War, it was predominantly the American crop development policy that was transferred to the newly independent former colonies and other developing countries, by means of American agricultural research aid programmes and by FAO seed dissemination activities. The introduction of the state-led crop development policy substantially changed the organization of agricultural production in developing countries. Governments regulated the seed market, actively promoted the substitution of landraces by new, scientifically bred plant varieties, and encouraged the introduction of agro-chemicals. These changes greatly facilitated the process of agro-industrialization. Agricultural regions of many developing countries increased their productivity and were able to export. In fact, they became incorporated into transnational production chains chiefly for the production of meat, fruit and flowers.

This section first examines the role of the FAO as a distributor of finished varieties and conservator of landraces. Then, we focus on the reasons for and the effect of the agricultural development aid by the U.S. government and some American philanthropic organizations. Under the Point-Four programme, the USA set up an extensive breeding programme on wheat and maize in Latin America, complemented by a plant conservation system. The complementarity was subsequently repeated in other international agricultural research centres (IARCs) in the 1960s and after. Finally, the involvement of the FAO and the USA in the industrialization of agriculture in the Third World initiated parallel efforts to set up a global plant conservation system. In the last section, we illustrate how this situation led to a conflict over the control of the global system.

4.5.1 The FAO's seed dissemination and exchange programme

The yield increases achieved by the American plant-breeding sector were substantial and widely known after the Second World War. They gave rise to an optimistic and assured FAO initiative, in the late 1950s, to re-organize the seed distribution system in developing countries in order to eradicate hunger. The FAO's seed programmes culminated in the World Seed Year, organized in 1961, a year that was dedicated to "the promotion of first class seed of superior crop and tree varieties" (FAO, 1960a:2). First class seed was considered to be higher yielding, disease and draught resistant, and adapted to mechanical harvesting (FAO, 1958b:4). The World Seed Year fell within the Freedom From Hunger Campaign and seems to have been fairly successful. The 1960 World Seed Campaign News, a journal that monitored the preparations and achievements of the World Seed Year, proudly reviews the many new seed certification bureaus in developing countries, established by new National Seed Campaign Committees with FAO support (FAO, 1960b:7). In the 1960s, the National Seed Campaign Committees were the primary actors in setting up a controlled distribution system of imported varieties in the Third World.

In order to stimulate the shipment of new varieties to developing countries, the FAO, in collaboration with the United Nations Educational Scientific and Cultural Organization (UNESCO), established a Gift Coupon Project. The Project supplied National Seed Campaign Committees with new varieties and agricultural equipment from industrialized countries. The Project reflected a rather supply-driven approach: Donor-organizations and charity groups bought coupons from UNESCO and offered these to the National Seed Campaign Committees.

The imported new varieties were deliberately considered to replace traditional seed stocks. The FAO's World Seed Campaign News (FAO, 1958b:5) mentions no fewer than 27 measures to bring "first class" seed to developing countries. The measures, for example, demanded "facilities for representatives of foreign breeders and seed firms [for] the introduction of high quality seed of approved varieties from abroad." Simultaneously, the Committees were to avoid the use of "inferior seed of unsuitable varieties" by local farmers (FAO, 1958b:5). The FAO's stance on the abandonment of local varieties is clearly stated in a text presented in 1958 by the FAO International Rice Commission:

> "Although in most countries ... rapid progress has been made in rice breeding, an excessively large number of rice varieties ... is still in cultivation in many countries. To increase the average yield of rice it is important that the lower yielding types be eliminated. ... Only a limited number of improved varieties should be included in a seed programme." (FAO, 1959a:9).

"First class seed" specifically included hybrid varieties. The 1962 newsletter of the FAO Freedom From Hunger Campaign reflects the FAO's expectations as regards hybrid varieties: "Enormous progress has been achieved in this field in the last four or five decades in the agriculturally advanced countries. A convincing example in recent times is the immense success of hybrid maize in the USA." (FAO, 1962:2). Hybridization was considered "today's most important type of breeding" (FAO, 1958b:6).

The FAO's role was limited to the dissemination of new plant varieties. The organization did not embark on large 'in house' crop development programmes. It had a mandate to deliver short-term food security, and the organization's budget - not more than US$ 80 million at the end of the 1960s (Talbot, 1990:29) - did not allow for such expensive initiatives. Also, the FAO's Crop Production and Improvement Branch merely functioned as a seed exchange service. Limited budgets and political strife between member countries also frustrated and prevented the establishment of Plant Exploration Centres for research on locally available landraces in Turkey, Argentina and Pakistan (Pistorius, 1997:17-18).

Although the FAO's direct role in crop development was limited, it did pave the way for the export of a state-led seed distribution system of finished varieties. The activities of the Crop Production and Improvement Branch in distributing unimproved material would in due course become even more important. As the Branch distributed and registered source and destination of tens of thousands of landraces and parental lines, it was gradually able to obtain a fairly accurate picture of the world's stock of landraces. This information was published in the FAO Plant Introduction Newsletter (FAO, 1957-1970) and the World Catalogues of Genetic Stocks (FAO, 1950a,b; 1959b). The newsletter and the World Catalogue gradually became key information sources for breeders worldwide.

It was with this global coverage in mind that an FAO expert team was commissioned to design a new policy for the conservation of landraces and wild relatives. In the early 1970s, the team, under the name of the Panel of Experts on Plant Exploitation and Introduction, came up with a proposal to set up a world-encompassing seed storage system. In line with Vavilov's broad scientific interest, each seedbank would maintain a broad 'generalist' conservation strategy, also covering many minor food crops of local importance (FAO, 1969; FAO, 1973). This high ambition soon aroused opposition from the other major player involved in exporting industrialized agriculture, the USA.

4.5.2 Sowing and collecting seed in Latin America

In a meeting with the Rockefeller Foundation about agriculture in Mexico in 1941, U.S. vice president Henry Wallace, the man who had been so instrumental in initiating the restructuring of American agriculture, noticed that: "In the field of agriculture ... there has been no long-term policy and there has been a lack of continuity in the programmes which have been adopted." (Wallace quoted in Lewontin, 1983:78). The policy Wallace had in mind went far beyond that of the FAO's seed campaign. Wallace envisaged a plan-wise industrialization of agriculture along the lines of the New Deal. The instruments of the associated crop development policy, later referred to as the Green Revolution, would be delivered by the U.S. government and private philanthropic organizations, notably the Rockefeller and Ford Foundations.

Green Revolution research strategies were carried by the ideal of the progressive entrepreneurial middle class-type farmer, superior to the peasant (Lewontin, 1983:27), and by the belief in the efficacy of modern, labour saving technology and the development of new high-yielding varieties. Most work was directed to the design of specific varieties that would facilitate the rationalization of farming. Breeding to reduce stem height of maize, rice and wheat was a key activity as shorter plants allowed for a better uptake of chemical fertilizer and more efficient mechanical harvesting. The other major task for plant breeders was the adaptation of new varieties to regional circumstances (day length, precipitation, and soil condition). Both types of work required a constant access to (local) landraces, which explains why seed conservation played a central role during the Green Revolution.

In 1943, the Mexican Department of Agriculture, with strong support of Wallace, embarked on a maize-breeding project that fell within the larger Mexican Agricultural Program (MAP) (Lewontin, 1983:77). While the Rockefeller Foundation had long-term experience with foreign agricultural aid, the MAP was its first long-term involvement in agricultural research. The MAP became "*the* method for effectively transferring technology from the developed to the less developed world ... through a complete partnership between the farmer, the state, modern industry, and modern science" (Perkins, 1997:92,117). The MAP would eventually evolve into the first IARC, the International Maize and Wheat Improvement Center (CIMMYT) in 1966 (Fitzgerald, 1994:73).

The significant contribution of the MAP/CIMMYT programmes to the increase in maize and wheat production in Mexico created concerns among the American agricultural establishment and conservative forces within the U.S. State Department (Oasa, 1981:106-112). Why did the U.S. government and the Rockefeller Foundation encourage the free international diffusion of advanced plant breeding technology, in which the country had a leading position? The answer must be sought in a mixture of American geopolitical concerns and (corporate) agricultural interests in genetic information.

The geopolitical concerns stemmed from the desire to create political stability on America's southern border. The economic depression of the 1930s and the outbreak of the Second World War had made Latin America receptive to populist and nationalist regimes. After 1943, the Roosevelt Administration responded to this situation with a non-interventionist Good Neighbour policy, which promoted American interests by other than military means. Like the New Deal, the Good Neighbour policy considered agriculture to be more than a residual sector in urban-industrial development. The objective was to foster rural social stability by transforming poor peasants into strong, progressive farmers. The translation of the Good Neighbour policy into technical agricultural assistance came in 1949 under the Point-Four programme of the Truman administration. The close collaboration between the U.S. government and agricultural experts of the Rockefeller Foundation allowed for a quick implementation of the Point-Four principles in Mexico (Lewontin, 1983:28-30,69; Perkins, 1997:108,117).

Mexico was the first candidate to demonstrate the new American foreign policy, and deserves closer attention for two reasons. First, the MAP integrated a long-term agricultural research strategy with conservation policies in a typically American way: centralized, well funded, and in accordance with the latest scientific insights. Second, the MAP's success (at least, in terms of production) created a format for the creation of many more IARCs in the 1960s and 1970s. It was during the MAP period that the agricultural research and conservation strategy of the Green Revolution first took shape.

The American initiatives were warmly welcomed by the Mexican elite. In the late 1930s and early 1940s, the Mexican agricultural sector was suffering from a crisis and parts of the rural population had become increasingly susceptible to socialist and fascist ideologies. Pressures from the rightist factions to curb the crisis by any means other than land reform induced the Camacho Administration (1941-1946) to find a 'technical fix' to increase agricultural production and defuse attacks on the power of the hacienda owners (Cotter, 1994:98). Another factor was the gradual replacement of maize and beans by wheat in the urban diet pattern. The growing preference for wheat created difficulties in the balance of payment, which made the emerging middle class favour domestically grown wheat. The proposals of the Rockefeller Foundation to set up an extensive wheat breeding programme thus fell in fertile ground, especially after the first large wheat crops in Mexico were hit by severe attacks of wheat rust (Perkins, 1997:111-112).

Next to geopolitical motives, the U.S. government had an interest in Mexico's rich diversity of genetic information. The MAP, and later CIMMYT, formed a perfect institutional base for the collection of genetic information of various crop plants, particularly maize. During the first few years, the MAP, the Rockefeller Foundation and the USDA extensively collected landraces wich were expected to be replaced by the new Green Revolution varieties. A Committee on the Preservation of Indigenous Strains of Maize was established, operating under the U.S. Agricultural Board of the National Academy of Sciences/National Research Council (NAS/NRC). The Committee cooperated with the Rockefeller Foundation and numerous Latin American governments and institutes (Brown, 1975:468; Dowswell, *et al.* 1996:58). In 1952, the Committee was joined by William Brown, research director of Pioneer Hi-Bred (USA), the world's largest maize breeding firm. According to a later vice-president for research of Pioneer, "Brown believed that for the future welfare of humanity, it was imperative to collect and preserve the fast-disappearing landraces and wild relatives."[19] Brown managed to get support within his own company to fund a special Latin American Maize Program (LAMP) which placed Pioneer at the centre of maize collection in Latin America (Cunningham, 1992:17-18).

Thus, the export of crop development technology from the USA to Mexico also set in motion a flow of genetic information in the reverse direction. This pattern would be repeated when in 1959 the Rockefeller and Ford Foundations and the USAID established the International Rice Research Institute (IRRI) in the Philippines (*cf.* Perlas and Vellvé, 1997:1-16). IRRI's seedbank has grown to become one of the best equipped and managed in the world. The state-led integration of plant breeding research and conservation activities was typical of American crop development policy, and was also subsequently transplanted to CIMMYT and IRRI. The 'model' policy consisted of three elements: (a) advanced research on a limited number of high-caloric food crops (root crops, beans), (b) a strict division of labour between the participating research institutions, and (c) one overarching bureaucratic system through which information on research results and collections could be distributed. As will be discussed in the next section, the replication of this model formed the basis for a global network of IARCs financed by the OECD countries, the Rockefeller and Ford Foundations, and the World Bank. The free flow of seed that gradually emerged within the global network would eventually raise tensions within the FAO.

At this point we may conclude that the successful replication of the American crop development policy in the Third World gave a strong impetus to the involvement of the USDA and American philanthropic organizations in the Green Revolution. Geopolitical interests muted protests from conservative forces in the American agricultural establishment that feared a diminishment of their export markets in the Third World. A hidden but no less important benefit was the function of the new international research centres (CIMMYT, IRRI) as transfer points in the global exchange of landraces and wild relatives. The control over these centres offered the USA a strong position in the global exchange of seed, plants and plant material. The long-term access to these sources of genetic information was considered paramount for the future of the American agricultural sector.

4.5.3 FAO and USA battle over global conservation

During the late 1960s and early 1970s, two parallel global conservation systems were emerging. In Rome, the FAO Panel of Experts was in the midst of designing a global conservation system that would extend the activities of the Crop Production and Improvement Branch in the dissemination of new plant varieties in developing countries (see section 4.5.1). Simultaneously, the Rockefeller and Ford

Foundations and, most important of all, the World Bank took an interest in extending the number of IARCs and associated seedbanks[20]. This parallel development raised the question of who was to control global conservation. Both parties had different but strong reasons to take the lead. The FAO's mandate to create food security made conservation of landraces a logical start for more research on food crops. The two Foundations and the World Bank, however, were specifically interested in strengthening the Green Revolution in areas susceptible to communism. By locating IARCs precisely in these areas, political turmoil could be prevented. Also, the 'seed flow' between the IARC collections could be of use for the agro-industry in the USA and other industrialized countries.

The World Bank and the Foundations had little interest in funding agricultural research and a global conservation system through the FAO. Firstly, the USA, and particularly the USDA had a difficult relationship with the FAO (Talbot, 1990:43-45), because the developing countries, due to the one-country-one-vote system, could easily create a majority. The atmosphere became worse when, in the early 1970s, the Group of 77 developing countries started to call for a New International Economic Order (NIEO). Increasingly tense NIEO discussions on access to and benefit sharing of 'global commons' (deep sea, space, Antarctica) in the mid-1970s indicated that the Group of 77 would sooner or later question the hitherto common practice of the free exchange of seed and plant material between the IARCs (Pistorius, 1995a:209-211). Secondly, the World Bank and the foundations understood that the export of the American crop development policy to developing countries would become very complex if the associated conservation strategies did fall into the hands of the FAO. The FAO funded national governments to set up their own national agricultural research systems (NARS). NARS, however, were not automatically supporters of the American crop development policy or the Green Revolution. In a later comment, World Bank vice-president Warren Baum explicitly states that "considerable doubt was expressed within the Bank about whether [OECD] countries would rush to borrow in support of national agricultural research programmes [the NARS]" (Baum, 1986:35).

The World Bank and the foundations thus had considerable interest in keeping the issue of global conservation away from the FAO arena. This was not an easy task since the FAO, as a UN organization, was a logical candidate to host a multilateral conservation system. Also, the FAO Crop Production and Improvement Branch had accumulated much experience in the exchange of seed among breeders of member countries. Moreover, the FAO Panel of Experts on Plant Exploitation and Introduction was well on its way to designing a global conservation system.

Although the World Bank group favoured the FAO Panel's ambition to think in global terms, the proposed broad, 'generalist' conservation strategy did not fit the more 'crop specific' model of the IARCs. The specialization of the IARC on one or two major food crops (CIMMYT on maize and wheat, IRRI on rice) was more in line with the crop development policies of the USA and other industrialized countries. From this point of view, the generalist approach as proposed by the FAO Panel, would have resulted in much broader and diverse collections than was necessary to pursue the Green Revolution. The Panel's idea to let each seedbank operate as an independent unit also failed to meet the World Bank's preference for a centrally controlled conservation system (Fowler and Mooney, 1990:152).

Nevertheless, the Panel's status as an FAO/UN expert group made it difficult to ignore their ideas. In 1972, the World Bank decided to invite the Panel to present its plans, in Beltsville (Maryland, USA). The outcome of this meeting would be decisive for the creation of the global conservation system. Because most Panel members had become sceptical about the FAO's ability to finance their plans, they agreed to integrate their ideas with those of the World Bank and the foundations (Pistorius, 1997:55). The resulting cooperation between the FAO and the financially much stronger World Bank reduced the status of the Panel to an advisory team, and eventually erased the FAO as a leading actor in the creation of the global conservation system.

Soon after the Beltsville meeting, the World Bank, and the Rockefeller and Ford Foundations accelerated their efforts to set up a network of research institutes and seedbanks. An important role was played by Robert McNamara who, during the first few years of his appointment as World Bank director, simultaneously ran the Ford Foundation. McNamara proposed to establish a "consortium" or "consultative group" of OECD donor countries, and large donor organizations: the Rockefeller and Ford Foundation, and the Inter-American Development Bank. The donors would meet periodically with agricultural experts from developing countries to develop policies for the consortium. The choice of a consortium, referred to as the Consultative Group on International Agricultural Research (CGIAR), served the World Bank's ambition to play a more prominent role in agricultural research (Baum, 1986:35-38). Besides, it allowed Canada, Germany, the United Kingdom, Japan and France to become more involved in the hitherto U.S. dominated long-term international agricultural research infrastructure.

The CGIAR had won a battle but not the war over the global conservation system. Due to its status as a non-governmental organization, a formal relationship with the FAO in conservation was still required in order to have unhindered access to seed stocks in developing countries. At the cost of another diplomatic fight that lasted throughout 1973, a rather peculiar compromise between the World Bank and the FAO was struck. It was agreed that the CGIAR would create and finance a new conservation institute. This institute would function independently from the FAO but, nevertheless, be located at FAO headquarters in Rome. The institute, named the International Board for Plant Genetic Resources (IBPGR), was established in 1974 and was entirely dedicated to the international exchange of seed. Under supervision of the IBPGR, the CGIAR conservation system grew into a global exchange system of seed and plant material, and one of the main institutional pillars supporting the Green Revolution.[21]

As a scientific, autonomous and politically 'neutral' organization, IBPGR aimed at operating within the FAO bureaucracy, but beyond the political reach of FAO member countries. It was this shielded position, in combination with the secure flow of finance from the World Bank/CGIAR headquarters, that allowed IBPGR to set up an international seed exchange network that went far beyond the CGIAR seedbanks. In 1984, only ten years after its establishment, the IBPGR network bound together about 40 national collections in 30 countries (FAO, 1985:2).

IBPGR, in spite of its modest financial resources, was soon able to play a primary role in the international exchange of seed and plant material. One of the main reasons was that IBPGR capitalized on existing national plant and seed stocks. Some of these collections dated back to colonial times, but most had been established shortly after the Second World War, to support the export-oriented agricultural policies of developing countries (Pistorius, 1997:1-3)[22]. Many of these collections suffered from neglect when national agricultural policies shifted away, either to bulk food production (wheat, rice), or to other export crops. IBPGR's modest financial and technical aid (often not more than several tens of thousands of dollars), frequently rescued national collections. The resulting durable contacts allowed IBPGR to set up a very cheap, but nevertheless global system in which national collections were gradually linked to those of the CGIAR.

The other pillar supporting IBPGR's key position in international conservation was its early development of crop descriptor lists to characterize germplasm accessions, some of them computerized, and the linkage of these lists with existing characterization methods used in OECD countries and CGIAR centres. The control over the design of these descriptor lists offered IBPGR the opportunity to design worldwide standard procedures for data collection and information exchange. Guidelines for the exchange of seed, plants and plant material, International Genebank Standards, and the development of more advanced conservation methods, all played a role in exchanging and conserving material around the world.

IBPGR's role in global conservation would nevertheless remain vulnerable. The quasi-independent status of IBPGR - being hosted by the FAO in Rome, but financed and controlled by the CGIAR - started to attract criticism from the Group of 77 only a few years after its establishment. IBPGR's staff knew that any political uproar over the exchange of seed and plant material would have severe repercussions for frequent users, particularly breeders in public and private institutes in industrialized countries and other CGIAR institutes. This explains IBPGR's minimal interference with national agricultural politics in developing countries, as well as its self-declared political 'neutrality'. Many developing countries accepted this policy as long as IBPGR and CGIAR policies indicated that the free exchange of seed would eventually contribute to food security. However, the status of the IBPGR vis-à-vis the FAO as a multilateral institution was one of the causes for the 'PGR conflict' in the early 1980s, which we describe in the first chapter of this book.

In our view, the global conservation system under guidance of the CGIAR is a true characteristic of the Second Agro-Food Order, despite the fact that the system has continued to function in the subsequent Agro-Food Order. An important reason is the main object of conservation. CGIAR's collections mainly consist of landraces, which were the main source of germplasm for breeders using conventional breeding techniques during the Second Agro-Food Order. As we show in chapter 5, conventional breeding is no longer the typical tool for crop development in the Third Agro-Food Order. Rapidly evolving genetic engineering techniques lead to new conservation strategies in which landraces no longer play a key role.

Summary

The FAO was the first international organization to set quality standards for seed in developing countries in the 1950s. Seed quality standards were meant to establish controllable seed distribution systems to secure the distribution of uniform and high yielding varieties. Although the FAO was instrumental in setting the new standards, its impact on research and conservation in developing countries remained limited. Partly, this was related to the limited economic resources of the FAO, partly also to the fact that uniformity and high yields were not criteria set by the receiving farmers in developing countries.

A second and more important factor of the limited success of the FAO was that the USA put most energy into an alternative route. Rather than seeking to transplant crop development policies through multilateral aid programmes, the U.S. government established long-term bilateral agricultural programmes in countries that served American geopolitical interests most closely. The clear and pragmatic structure of the American conservation system, which was highly centralized, crop specific, and closely connected to crop development projects, allowed for a quick replication in developing countries. This asset, combined with strong capital injections from the World Bank, the Rockefeller and Ford Foundations, created the basis of an international network of agricultural research stations under CGIAR. The establishment of the IBPGR as a 'traffic agent' for the 'seed-flows' within the CGIAR system completed the system.

Conclusions

In this chapter we have explained how, in advanced capitalist countries, the state became the key organizer of crop development during the Second Agro-Food Order. The main goals for crop development were set by the strong governmental emphasis on productivity increase, while most governments actively promoted private investment in plant breeding. Apart from an intensification of seed market controls, which had begun in the First Agro-Food Order, the government initiatives in crop development concentrated on two areas:

Funding basic and pre-competitive breeding research

Governments in most advanced countries invested substantially in basic research, such as genetics, because this was beyond the financial reach of the generally small, family-owned breeding firms. Particularly the American private breeders have relied heavily on the parental lines developed by public institutes. Undoubtedly, the government support for breeding research greatly contributed to the tremendous yield increases of the main crops. A similar picture can be offered for conservation in the USA. The establishment of a large-scale and centrally organized conservation network could never have been carried out by the private breeding firms. The public investments prepared the area of crop development for private investment. In other words, public investment in breeding conservation helped to make plant breeding a more controllable industrial sector, attractive to private capital.

Reducing unauthorized propagation

Another way to promote private investment in crop development was the assistance offered to private breeders in reducing unauthorized propagation of their varieties. The hybrid varieties, which were developed with strong support from the USDA, effectively diminished this problem for plant breeders involved in various crops. For crops in which hybrid varieties were not developed the system of plant breeders' rights (PBR) was created by governments in Western Europe. PBR legislation offered a legal means to reduce unauthorized propagation for a broader range of crops, but was less effective than hybridization and prone to political decision making.

In an effort to eradicate hunger and prevent communist influence in the newly independent countries, the state-led crop development policy was expanded to the Third World, on the initiative of the U.S. government and the FAO. Seed market regulation, as it had evolved in Europe and the USA since the end of the 19th century, was introduced in developing countries, and so was the policy of public investment in breeding research and in conservation. The diffusion of hybrid varieties sufficed as protection against unauthorized propagation in developing countries during the Second Agro-Food Order. Domestic private investment in crop development was marginal if it existed at all, while the Third World was generally not an attractive market for private European and American breeders. Most foreign varieties that were sold involved hybrids.

The central role of the state in the organization of crop development during the Second Agro-Food Order was not a predetermined and uniform phenomenon. Governments in the various countries entered the area of agriculture and crop development because they were supported by their national political constituency. State interventions in crop development were generally shaped by political disputes. The main controversies in crop development during the Second Agro-Food Order were over the legal protection of plant varieties and on plant conservation.

While governmental support for creating hybrid varieties did not provoke major opposition, legal protection for plant varieties was controversial, particularly in Europe. Throughout the entire Agro-Food Order there was a tug-of-war between private plant breeders, the general industry sector, and agricultural authorities over the most appropriate legal way of helping private breeders to reduce unauthorized propagation of their varieties. The result was a compromise. Although PBR legislation provided far less protection than the breeders had envisaged, it was accepted, since the Western European and U.S. governmental authorities were not willing to offer more, while industry did not want patents for plants to affect 'its' patent system.

The second controversial area was conservation. In the USA plant conservation has traditionally been centrally organized by the state. In Western Europe, however, a tradition existed that breeders, private and public alike, maintained their own working collections. This decentralized, or rather fragmented, manner of plant conservation led to several attempts to organize conservation at a common, European level. These attempts came to nothing, however. Intra-European controversies, even among the members of the EEC who had established a Common Agricultural Policy, prevented working collections and seedbanks from being organized more centrally. The lack of political support for a central conservation approach presumably results from the availability of landraces of most commercial crops in or near Europe.

Another controversy that influenced governmental intervention in conservation, this time at an international level, was the dispute over the control of the global conservation system. While developing countries favoured worldwide conservation through the FAO, this was not acceptable for the USDA and for some organizations related to the U.S. government, such as the World Bank, the CGIAR, and the major American philanthropic foundations. The global conservation system under guidance of the U.S. government not only served the exchange of seed and plant material among developing countries, but also between developing and industrialized countries.

Notes

1 Figures on public agricultural research expenditures in the main industrialized countries during the post-war period are hardly available. An authoritive research group in this area reports a "massive wave of public investment in science and technology research in general and agricultural research in particular" after the end of the Second World War. Only for a few countries is more specific data given. Between 1950 and 1970, in real terms public agricultural R&D expenditures increased at annual average growth rates of about 7% in Australia, about 7.4% in the United Kingdom, and about 2.8% in the USA. For New Zealand and The Netherlands similar data are not available, but for these countries substantially increased public investments in agricultural R&D in the same period have also been reported (Alston *et al.*, 1998:55).

2 It should be noted here that a calculation of the 'net' contribution of breeding work to yield increase (or "genetic contribution") is very difficult to assess because of the interplay between the new genetic constitution of a variety and other production factors. The yield increases can partly be attributed to the improved nitrogen uptake of plants.

3 In the 1950s American farm employment was reduced by 28% in the 1950s and 37% in the 1960s (Cochrane and Ryan, 1976:13). The reduction of farm employment in the initial six EEC countries in the 1960s was 38% (Commission of the EEC, 1988).

4 The national plant variety catalogue (also referred to as Variety Register or Variety List) could vary from a simple registration merely of variety names, to a register that included judgements about the

variety's value for cultivation and use, based on tests carried out by agricultural authorities.

[5] The Decree put the seed market under strict governmental control, but this was unrelated to the fact that at that time the country was occupied and ruled by the German Nazi regime (Addens, 1952:185-187).

[6] In 1948, seed firms from 13 countries participated in ASSINSEL: Belgium, Czechoslovakia, Denmark, France, The Netherlands and Sweden. Representatives from Austria, Great Britain, Hungary, Italy, Luxembourg, Spain and the United States had an observer status (ASSINSEL, 1948).

[7] The German seed industry began to participate in ASSINSEL in 1950 (ASSINSEL, 1950:4).

[8] While explaining the working of the Dutch breeders' right system during the 1950 ASSINSEL Council meeting, a representative of the Dutch horticultural industry, Mr. Ruys, argued that the Dutch law "is only adapted to the interior Dutch regulations; it is a state-organism which functions in a very rigorous way." (ASSINSEL, 1949b:10).

[9] The Danish, French and Italian lawyers preferred patent protection for plant varieties. The groups from Luxembourg, The Netherlands, Switzerland and the United Kingdom advocated a specific protection system for plant varieties. The German group was in favour of a mixed system: patents for major breakthroughs in plant breeding and special protection for "ordinary" new varieties (Heitz, 1987:78).

[10] Art. 2. "the Contracting States *shall not be bound to provide for the grant of patents* in respect of ... plant or animal varieties or essentially biological processes for the production of plants and animals; this provision does not apply to micro-biological processes and the products thereof." (italics added).

[11] Article 52. "European *patents shall not be granted* in respect of: ... (b) plant or animal varieties or essentially biological processes for the production of plants and animals; this provision does not apply to microbiological processes or the products thereof." (italics added).

[12] From the historical review of the conference presented by Heitz (1987:82-84) it can be concluded that 13 of the 16 persons who "may be considered the founders of the UPOV Convention" represented a national Department of Agriculture or an agricultural research institute.

[13] The concern by the U.S. seed industry was well expressed by John Carew, of the Department of Horticulture at Michican State University, who noted in 1964: "Many commercial seed firms see a need for some form of breeder protection. But an overwhelming majority strongly oppose compulsory registration, particularly the idea of any group of men, government or industry appointed, making arbitrary decisions on the distinctness, homogeneity, or stability of any variety" (Carew, 1964:38).

[14] Plant Breeders' Rights (PBR) protection seems to be European wording of what in the USA usually is called Plant Variety Protection (PVP) rights. As both terms have essentially the same meaning they will be used interchangeably in this publication.

[15] Doyle (1985:63-65) and Fowler (1994:115-118) describe in great detail the struggle about PVP that broke out among the various departments involved in the White House review process only a few days before the President had to act on the bill. The Bureau of the Budget and the President's special assistant for consumer affairs advised the President to veto the bill because of concern for increased market power of seed producers, and the lack of evidence that PVP would stimulate plant breeding. Due to a not entirely open political lobby, the recommendation to veto was reversed at the very last minute.

[16] The growing technical opportunities for industrial processing of crops after the Second World War induced plant breeders and seedbank curators to refer to landraces and wild relatives as "germplasm resources" or "germplasm" (*cf.* Hodgson, 1961). The term "germplasm" is confusing, however. It is commonly used in biological science, but the term has never been defined properly. The term basically refers to the physical basis of heredity, such as chromosomes and genes, cytoplasm and protoplasm. (*cf.* Witt, 1985:7-11; Somani, 1992:81).

[17] The need for the establishment of temperate seed collections in Europe was already clearly voiced

during the 1936 International Congress of Plant breeders, where chairman C. Broekema from the Netherlands remarked: "There should be in every country extensive living collections of varieties at the disposal of breeders. ... This point is urgent since Russian expeditions have discovered rich centres of diversity in remote parts of the world, and everybody is doing his best to gather some of the genotypes from these centers. ... An international list could be made, mentioning not only the varieties used in praxis, but also describing interesting biotypes ..." (Broekema quoted in Sirks, 1936:42)

[18] Point-Four refers to the fourth issue of Truman's inaugural speech of 1949, which stressed the need for development aid for developing countries.

[19] Information derived from a personal communication with D.N. Duvick, former vice-president of research, Pioneer Hi-Bred, August 1998. According to Duvick, Brown's initiatives in collecting maize landraces went far beyond the corporate interests of Pioneer Hi-Bred. Initially he attempted to get the American maize seed industry collectively to fund conservation facilities in Latin America to store maize landraces. The seed companies would not contribute. Nevertheless, Brown talked his company into donating the entire amount, and personally got the programme started. Once he left the management of the company, funds for the project were terminated.

[20] After CIMMYT and IRRI four other centres were built: the International Center for Tropical Agriculture (CIAT) in Colombia in 1967, the International Institute of Tropical Agriculture (IITA) in Nigeria in 1968, the West Africa Rice Development Association (WARDA) in Ivory Coast in 1971, and the International Potato Center (CIP) in Peru in 1971. By the early 1990s, the total number of centres had risen to 16, spread over five continents.

[21] Very likely, the conservation network of the COMECON countries, including the VIR or the French conservation system maintained by the French *Office de la Recherche Scientifique et Technique d'Outre-Mer* (ORSTOM) were, in terms of size, comparable with the CGIAR conservation system. The socio-economic impact of these systems on crop development and agriculture in developing countries, however, was significantly smaller. COMECON's conservation system mainly serviced countries in the communist sphere of influence, while the collections of ORSTOM were (and still are) geared to consolidating the export of cash crops (notably fruit, coffee and tea) from former colonies (personal communication Ph. Auriau, INRA, 1997).

[22] Particularly Colombia's national collection of native potatoes, grasses, pasture and legumes, Costa Rica's collection of cocoa, and India's collection of pulses, crucifers, grasses and forage legumes had a very high international reputation in the 1950s and 1960s (Pistorius, 1997:2).

Chapter 5

The Third Agro-Food Order (1980s-) Crop Development Conglomerates Feeding the Eight Billion

"Plant genetic resources have a key role in world trade". This was one of the statements distributed by the CGIAR on the eve of the United Nations Conference on Environment and Development (UNCED) in 1992. The 'fact sheet' on which it was printed contained a clear message: non-traditional exports based on local genetic resources formed a major window of opportunity for developing countries to explore new niche markets: "An example is the *Pitahaya*, a fruit-bearing cactus native to central Colombia which ... is now a thriving export crop, mainly to Japan." (CGIAR, 1991:1)

That plant genetic resources play a key role in world trade is undeniable. But the idea that local varieties allow developing countries to explore niche markets in non-traditional export markets is a mistake, as is CGIAR's *Pitahaya* story. This chapter examines the main reasons why it is not Colombia, or other developing countries, but the private crop development industry based in OECD countries that will be able to pick the fruits of the non-traditional export species.

Since the mid-1990s, chemical, pharmaceutical, and food companies have been making unprecedented take-over bids for plant breeding and genetic engineering firms, to secure their involvement in the crop development industry of the 21st century. A steep rise of private investment in crop development has been accompanied by a worldwide adoption of neo-liberal economic polices and a reduced involvement of governments in agriculture. These tendencies have facilitated the emergence of large industrial conglomerates that possess the financial and technological capacity to exploit plants and other organisms at the molecular level for crop development. They place OECD countries at the core of a new, truly global market of plant varieties. In section 5.1 we examine the specific role that can be attributed to the private crop development industry in shaping the new international division of labour in agriculture and the corresponding regulation that characterize the Third Agro-Food Order.

The rise of an influential private crop development industry has significant consequences for conservation policies. During the Second Agro-Food Order, plant collection and conservation focused on landraces as main resources for plant breeding. Genetic engineers seem no longer to focus exclusively on these plant types. Section 5.2 explores how deoxyribonucleic acid (DNA) data banks are becoming the industry's primary sources of genetic information, and what implications this has for the present seedbanks. Subsequently, ample attention is devoted to the suppliers of the DNA useful to the industry. In the tropical forests and in farmers' fields in developing countries a large part of the world's biological organisms, carriers of potentially useful DNA can be found. Can developing countries benefit from their supplier function?

The final section deals with the changes in plant-related intellectual property protection which the world has witnessed since the 1980s. During the Second Agro-Food Order, breeders aimed at a reduction of unauthorized propagation of their plant varieties. The large industrial involvement in plant breeding and the opportunities to 'engineer' a new plant have made the protection mechanisms, as established during the previous Agro-Food Order, ineffective. Apart from propagation, genetic engineering corporations have a strong interest in controlling the exploitation of genetic information in general. The discussion on the main changes in, and controversies on, plant-related IPR protection in the USA and Europe is followed by an exploration of the effects of the new protection standard on breeders and farmers in developing countries.

5.1 Towards an international division of labour in crop development

During the First and Second Agro-Food Orders, the process of industrialization of agricultural production relied mainly on the refinement of mechanical tools (tractors and harvesting machinery) and chemical inputs (insecticides, pesticides, fertilizers). In the 1970s, however, the impact of these production factors on agricultural output started to level off, and in some cases even to decline. It is against this background that we position the emergence of the biotechnology industry in the 1980s. What agro-industrial interests did this new industry meet? What impact did it have on the traditional dominance of public organizations in crop development? The sheer size and future scenarios of the new crop development giants forecast changes in the organization of crop development and the international division of labour in agriculture. The impact of the new industrial 'crop development conglomerates' on agricultural production in the Third Agro-Food Order is the subject of this section.

5.1.1 The emergence of the Third Agro-Food Order

As we have explained earlier, the Agro-Food Orders comprise a particular international division of labour in agriculture, facilitated by a specific regulatory framework. When the regulation no longer fits the division of labour, the 'order' enters a crisis, in which the foundations are laid for a new Agro-Food Order. In this way, the First Agro-Food Order can be considered as the outcome of the agricultural crisis of the 1870s. The vulnerability to cheap agricultural imports and low prices pressed governments in all capitalist core countries to undertake initiatives for the improvement of national agricultural competitiveness. These initiatives included a regulation of the seed market, and support for elementary crop development research. The Second Agro-Food Order arose from the Great Depression during the 1930s. The crisis gave rise to the gradual design of a comprehensive strategy for the industrialization of agriculture, adopted in all OECD countries after the Second World War. Instead of just interfering marginally, as they did during the First Agro-Food Order, governments grew into central organizers of agriculture and crop development.

The Third Agro-Food Order emerged from the agricultural crisis of the 1980s, which resulted from changes at both the production and regulation level, particularly triggered by (a) the rise of oil prices and (b) vast farm support outlays.

(a) Rising costs of agro-chemicals

The prevailing crop development policy during the Second Agro-Food Order relied heavily on fuel-based inputs. The rise in agricultural productivity had been strongly related to the increased use of chemical fertilizer, pesticides and herbicides. By the early 1970s, American farmers were burning about two and a half calories of fossil fuel to produce one calorie of harvested food (Goodman *et al.*, 1987:101). This high dependence on fuel became a matter of immediate political concern in 1973, when the OPEC cartel started to raise oil prices. Concern increased when it was revealed that the incremental effect of fuel-based fertilizer had decreased in American agriculture (Brown, 1985:34; McMullen, 1987:25).

There were also indications that the effectiveness of insecticides and pesticides was being undermined. A rapid spread of the Greenbug pest caused an estimated US$ 100 million loss to the U.S. sorghum crop in 1968. Two years later a corn leaf blight epidemic shocked the agricultural establishment as it wiped out 15% of the total U.S. maize harvest (Doyle, 1985:1-15). The declining effect of agro-chemicals was grist to the mill of environmentalists. The contamination caused by chemicals, particularly DDT, had become a key environmental issue since Rachel Carson published her book *Silent Spring*, in 1962. The criticism of high-input agriculture and the role of agro-chemicals transcended the U.S. borders and reached a temporary climax at the international level in 1972 during the United Nations Conference on the Human Environment (UNCHE) in Stockholm (McCormick, 1989:56,97). The conference confirmed the growing concern about the human mismagagement of the earth and, for the first time, attempted to encourage national governments to make the environment a policy issue. The recession of the 1970s would postpone governmental initiatives for more than a decade, however.

(b) Surplus production and public spending

The agricultural production strategy that had prevailed throughout the OECD countries during the Second Agro-Food Order had been very successful in terms of production. However, as national producers were largely shielded from the international market, a mechanism to control total farm output failed. Internal agricultural prices remained relatively high and delayed the reduction of the number of farms. The upward spiral of productivity increase and farm support could remain intact as long as foreign outlets for the surpluses were found. Even when industrial production in the USA and Europe entered a crisis in the 1970s, and the backbone of the post-war economic regulation - the Bretton Woods system - was shaken to its roots, agriculture flourished. Grain purchases by the USSR, the increasing food imports by the newly industrializing and oil producing countries, and acute food shortages in other parts of the world promoted agricultural production and export (Kenney *et al.*, 1991:182). Between 1970 and 1981 the American and European Community (EC) agricultural exports rose respectively to 30 and 20% of total world production (Newman *et al.*, 1987). Only after 1981, during the severe international economic recession, did agricultural exports stagnate. To a certain extent, the EC, Argentina and Brazil could still expand their market shares at the expense of the American exporters, who were suffering from the high value of the dollar. Overall, however, the growth in demand stagnated, causing giant food surpluses on both sides of the Atlantic. Total government outlays in the USA and EC for storage, income and price support amounted to more than US$ 50 billion in 1987 alone (Newman *et al.*, 1987:53).

The regulatory response to the agricultural crisis of the 1980s was twofold. In the first place a solution had to be found for the burdensome transfer from taxpayers and consumers to the agro-food sector. The massive governmental support for the national farm sector met with increasing political opposition. Moreover, the protectionist measures were challenged internationally as shown by the trade disputes between the USA, the EC, and the Cairns Group of Fair Trading Nations representing small and medium-sized agricultural exporters. While embracing an overall neo-liberal economic policy, governments in all OECD countries have responded by diminishing their central and protective role in agriculture, implying a shift towards a market-led agricultural production strategy. Price support measures tend to be replaced by direct farm income support, while total outlays for agriculture are declining. Nevertheless, total governmental farm support in OECD countries is still considerable and estimated at around US$ 166 billion for 1996 (Viatte and Schmidhuber, 1998:159).

A new crop development policy predominantly based on genetic engineering can be considered as the second major strategy to resolve the crisis and speed up industrialization of agriculture. In the 1980s, public investment in biotechnology became a recognized public tool to support advances in molecular biology in all main OECD countries (Roobeek, 1990:114-124). Such basic research had an important role to play in the alleviation of the agricultural crisis. It should help to make agriculture less dependent on fuel-based chemicals and more environmentally sound, and should reduce surpluses by offering new opportunities for agricultural diversification. In short, biotechnology was considered to be indispensable to increase the national competitiveness in crop improvement and agro-food production.

The interest in biotechnology arose in the 1970s, when a series of discoveries in molecular biology enabled the unravelling and modification of the hereditary mechanism of organisms at the genetic level. The workings of the hereditary mechanism was first explained in 1953, when the biologists Crick and Watson were able to unravel the structure of the DNA molecule. This discovery allowed scientists to understand how DNA is able to store and pass on hereditary information through combinations of four different bases (named guanine, cytosine, thymine and adenine). From this moment on, sequences of DNA base pairs became the main unit of genetic information in science. It would take until the 1970s, however, before 'genetic engineering' techniques were developed which enabled the modification of the genome and the transfer of the re-combined DNA to progeny.

The advances in genetics were quickly applied to crop development. Until the 1970s, specific plant characteristics could only be transferred from one generation to the other by means of crossbreeding techniques. These are based on a process in which the plant's own reproduction organs are the main tools. Genetic engineering techniques offered breeders the opportunity to pass on characteristics to the progeny under industrial, controlled circumstances through the isolation, recombination, and transfer of genetic information. These processes, also referred to as 'molecular methods', can be time saving as they are more accurate than crossbreeding. Molecular methods are based on genetic sequences which control predictable biochemical functions.[1] Genetic engineering also offers the opportunity to reduce the 'species barrier', the natural barrier that prevented 'traditional' breeding techniques to make crossings between different species. DNA from, for example, bacteria can now be transferred to plants. Genetic engineering thus gave crop development a series of new tools that had the potential to help overcome the major causes of the agricultural crisis: it offered technological opportunities for better control of the genetic make-up and characteristics of plant varieties, in order to diversify the use of surplus crops, and reduce energy consumption and environmental contamination.

The diversification objective initiated the generation of 'value-added' or 'speciality' crops. These predominantly comprise transgenic plant varieties designed for the production of specific industrial substances. The focus is on bulk crops, such as maize, soyabean, sunflower, and rapeseed (Duvick, 1996:16). Examples of the 'value-added' crops include Dupont's 'Optimum high-oil corn', which is claimed to contain nearly twice as much oil as traditional maize. Limagrain has developed tobacco plants that produce human haemoglobin and gastric lipase. Limagrain also directs its value-added programme to "nutraceuticals", medicinal compounds produced in vegetables to be used in animal feed, starch, oil, etc. (Freiberg, 1997a:8). Agracetus has produced transgenic cotton varieties with specific fibre characteristics, such as a blue colour and the ability to absorb water. Calgene markets transgenic oilseed rape with an enhanced lauric acid content, which produces rapeseed useful for the soap industry (Fitzpatrick and Scarth, 1998). The Centre for Plant Breeding and Reproduction Research in the Netherlands has developed the 'Fructanbeet', a sugarbeet producing high value fructanes instead of sugar (CPRO, 1998). Research investment in speciality crop production is highest in the USA. American chief executive officers (CEOs) of major crop development corporations predict that in 10 to 15 years half the U.S. grain will be grown for specific uses, as a value-added crop (Freiberg, 1997b:6).

Genetic engineering also offered inroads to reduce the use of fuel-based agro-chemicals. The idea is that new plant varieties are genetically modified to increase their resistance to pests and diseases and to make agro-chemicals superfluous. The first results, however, include plant varieties that seem to have the opposite effect. Cotton and soyabean varieties have been developed that are tolerant to glyphosate, an active compound that is the basis of various herbicides. It has been claimed that herbicide-tolerant varieties enable a more selective use of herbicides. The advantage for the industry itself is that the herbicide-tolerant varieties compensate losses on the agro-chemical market, while herbicide tolerance extends and adds value to the existing seed market (Bijman, 1998).

In sum, the prevailing regulation that seems to characterize the emerging Third Agro-Food Order is based on an agricultural production strategy in which the national agricultural sectors are less protected from the international markets than during the previous Agro-Food Order. The new crop development strategy focuses on genetic engineering, which offers the prospect of achieving more control over the hereditary mechanisms of plants, and thus the properties of new plant varieties.

5.1.2 The new international division of labour in agriculture

At the end of the Second Agro-Food Order, the international division of labour in agriculture underwent modifications, but was not disrupted. The USA increasingly had to share with the EC its position as the world's major food supplier, and rising production in developing countries prevented the USA and the EC from exporting all of their growing agricultural surpluses. Nevertheless, the overall position of OECD countries in the international division of labour in agriculture has been strengthened in the past decades. Between 1970 and 1996, the OECD share in the volume of world exports of cereals rose from 73 to 82% (Figure 5.1).[2] The USA has remained the world's major exporter of major commercial crops, including maize, soyabean and wheat (Figures 5.2 to 5.4). Despite increases in production of staple crops in Africa, Latin America and Asia, the share of these regions in the world imports has increased considerably (Figures 5.6 to 5.9). Together they account for nearly 60% of the volume of world cereal imports .

In the past decades, many initiatives were taken in developing countries to stimulate the export of non-traditional commodities (NTCs) such as new species of fruit, vegetables and cut flowers (Friedland, 1994; Thrupp, 1995). Although several developing countries have become successful NTCs exporters[3], the main beneficiaries of the tripling of the world market for fruit and vegetables have been the OECD countries. Between 1970 and 1996 their share in the volume of world exports expanded from 40 to 50%, while their share in world imports is declining (Figure 5.5 and 5.10).

We can assume that the growing share of OECD countries in the world agricultural market will persist. The already high level of industrialization of agriculture in these countries offers them a comparative advantage, particularly in cereal production. Large investments in crop development are likely to strengthen further their position in the international division of labour in agriculture. Most innovations in crop development are applied to temperate grains and temperate and subtropical vegetables grown in OECD countries.

In addition to their prominent position in the export of major agricultural crops, OECD countries are becoming the world's major suppliers of plant varieties. As we have indicated in earlier chapters, commercial seed has been traded internationally throughout the 20th century, but sales have mainly been confined to what now are the OECD countries. Still in 1990, over 90% of the world market for commercial seed, worth US$ 28 billion, was accounted for by these countries[4] (Figure 5.11) (Heijbroek *et al.*, 1996:17; Le Buanec, 1996:409).

Seed demand in OECD countries is diminishing, however. Between 1980 and 1994, the market for seed of arable crops stabilized in North America and decreased by 17% in Europe, mainly due to the decline of land under cultivation, and lower agricultural prices. The resulting intensification of competition in the seed market forces the crop development industry to explore new markets in developing countries where seed demand is expanding. Seed demand of arable crops increased by more than 30% in Asia between 1980 and 1994 and almost tripled in Africa in the same period (Heijbroek *et al.*, 1996:17).

The growing demand for seed of advanced varieties in developing countries has various causes. Population growth urges farmers to use ever higher-yielding varieties, or varieties that are adapted to marginal lands. As the major part of this growth occurs in urban areas, the dietary patterns of cities have become important factors in overall demand. Fast growing consumption of pasta, bread, and meat in cities has increased the demand for wheat and soyabean varieties. Imported and adapted plant varieties are also required to reduce the dependence on cereal imports, or to expand agricultural exports.

In developing countries, it is often imperative for agriculture to grow modern plant varieties, bred in OECD countries, in order to remain internationally competitive. These varieties are normally highly productive, provided that they are grown under specific conditions; furthermore, they meet industrial processing requirements, international quality standards, and the preferences of the OECD country consumers. In the developing countries themselves, the opportunities to create technologically advanced varieties have declined. Government supported research centres are still the main suppliers of new varieties in these countries, but public

Figure 5.1/5.10 Regional shares in world exports and imports of cereals, fruit and vegetables in 1970 and 1996*, in percentage (based on FAOSTAT, 1998)

Figure 5.1 Export shares: cereals

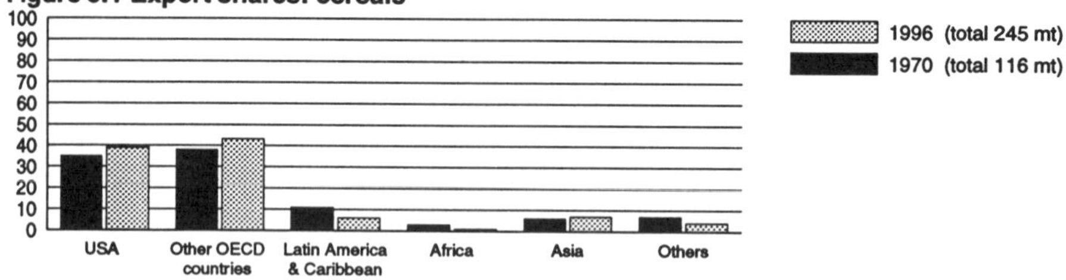

Figure 5.2 Export shares: wheat

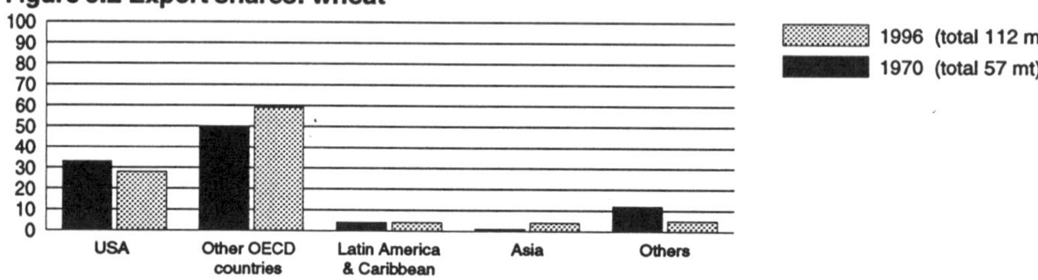

Figure 5.3 Export shares: maize

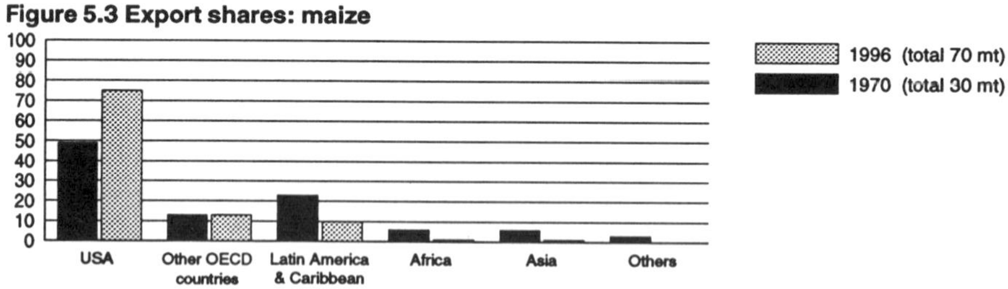

Figure 5.4 Export shares: soyabean

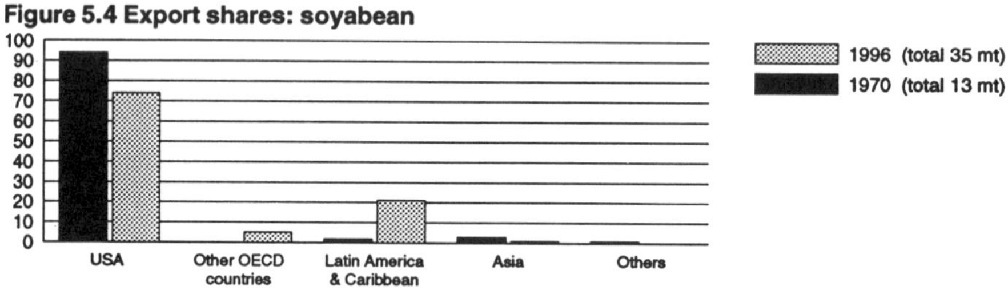

Figure 5.5 Export shares: fruit and vegetables

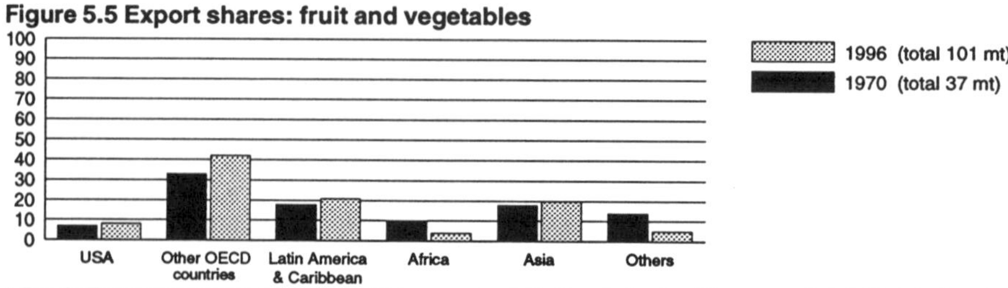

* Data for Eastern Europe and the former Soviet Union are not available. Data for Israel and Japan are not included in Asia. Japan is included in OECD. Data for Mexico and Turkey are not included in OECD but in Latin America and Asia respectively.

Figure 5.1/5.10 Continued

Figure 5.6 Import shares: cereals

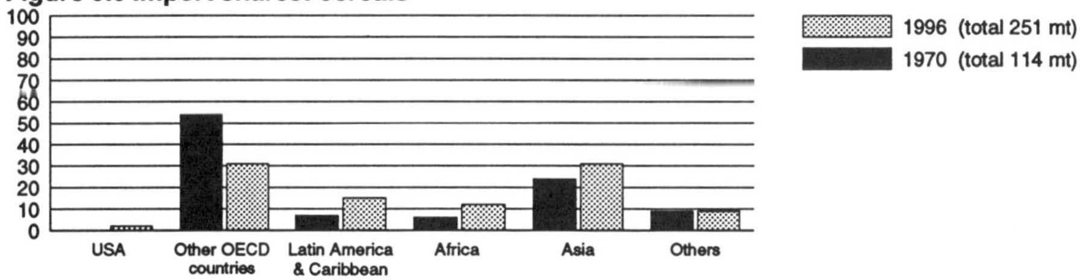

Figure 5.7 Import shares: wheat

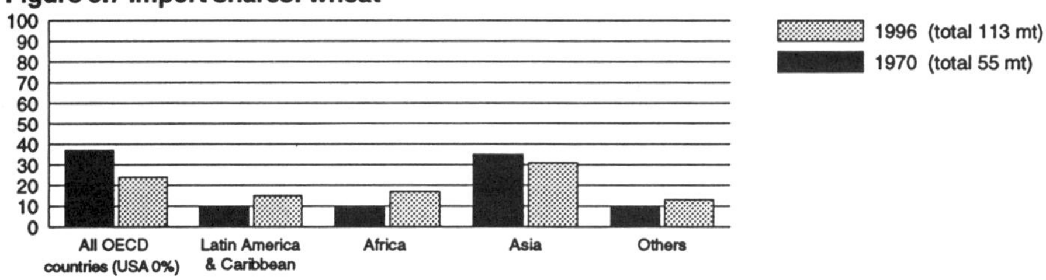

Figure 5.8 Import shares: maize

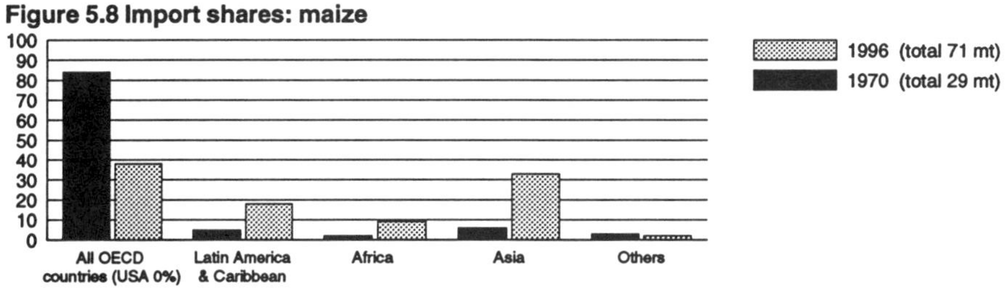

Figure 5.9 Import shares: soyabean

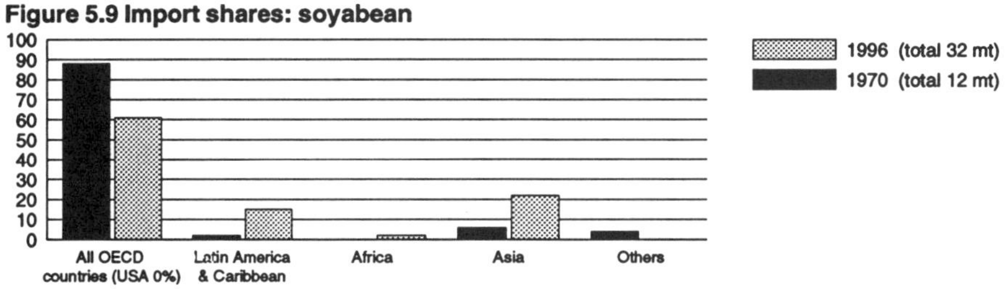

Figure 5.10 Import shares: fruit and vegetables

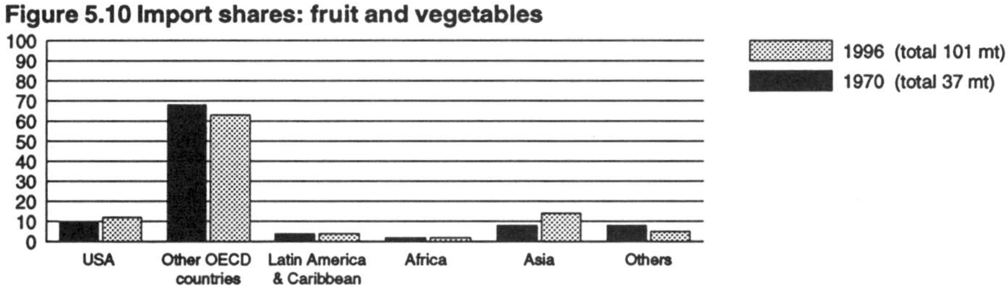

Figure 5.11 Shares in the world agricultural seed market by country (1994)

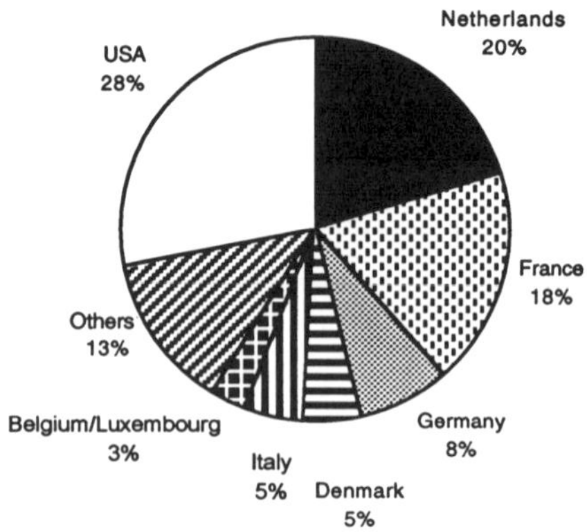

(Source: Heijbroek *et al.*, 1996:27)

plant breeding expenditure is under pressure. Whereas public agricultural R&D budgets continue to expand in most regions, the annual growth rates are declining (Figure 5.12). In Latin America, public expenditure on agricultural research has been reduced in absolute terms. The budgets of the national agricultural research institutes in 14 South and Central American countries were reduced by 13% on average, in the period from 1981 to 1993 (Echeverría et al., 1996:3). Most governments of developing countries are reducing their involvement in crop development because of structural adjustment programmes, while foreign donor support for plant breeding is diminishing (Tripp, 1997:25).

Figure 5.12 Annual growth of rates of public agricultural R&D spending in 22 OECD countries and 131 developing countries, in percentage (1971-1991)

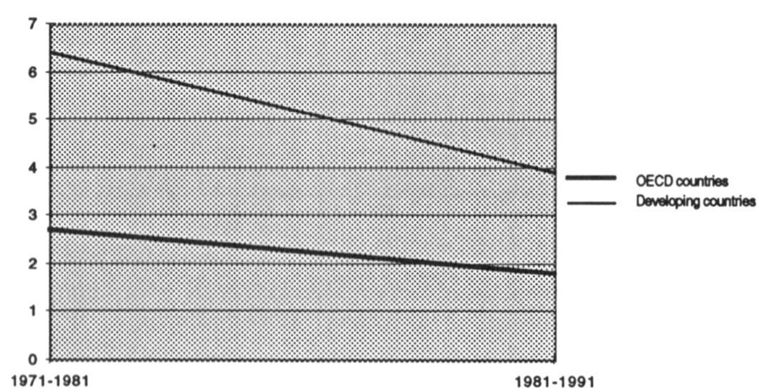

(Data derived from Alston *et al.*, 1998:59)

In developing countries the reduction of public research budgets is generally not compensated by increases in domestic private investment; private breeding, for example, is something that has only begun to emerge in the 1990s. The domestic firms often rely on the transnational crop development industry for the purchase of parental lines of hybrid maize, soyabean, fruit, vegetables and ornamental plants. Most firms concentrate on the adaptation of imported varieties to local climatic conditions and the sale of propagative material on the national market.

In sum, between 1970 and 1996 OECD countries strengthened their position on the world markets for staple crops and non-traditional commodities. Developing countries remained major importers of cereal crops, and overall they benefited little from the emerging markets for non-traditional commodities. The position of OECD countries in the international agricultural division of labour has been further strengthened by their new role as global suppliers of new plant varieties. The increasing requirements of new plant varieties, in combination with stagnant public investment in agricultural research, induce commercial agriculture in developing countries to rely on varieties produced in, or by companies based in OECD countries. Thus the Third Agro-Food Order has not only strengthened the existing international division of labour in agricultural production, but has also brought forth a division of labour in a subsector of agriculture: crop development.

5.1.3 The changing organization of crop development

The capacity of OECD countries to produce important plant varieties for diverging regions of the world seems very much to be determined by the changing organization of crop development. Whilst governments used to be the main organizers of crop development, there is a tendency for private industry to replace the national government in that role during the Third Agro-Food Order. This is a new phenomenon. Until the 1970s, large industrial corporations were involved in the production of agro-chemicals, but not in plant breeding. The lack of effective technical control over the plant's hereditary mechanism rendered plant breeding too risky a business for large industrial capital investment. This situation has changed since the 1970s, basically because of two mutually re-enforcing tendencies.

The first involved a relative decline in governmental expenditure in agricultural R&D, as part of an emerging market-led production strategy. Governments in both OECD and developing countries have started trimming their support for agricultural R&D[5] since the 1980s. Public agricultural R&D expenditure generally still grew, but at lower rates than they did during the Second Agro-Food Order. Overall in OECD countries, the annual growth rate of agricultural R&D dropped by a third, from 2.7% in the 1970s to 1.7% in the 1980s. In developing countries, as mentioned above, the decline was sharper, from 6.4 to 3.9% (Figure 5.12). In addition the public agricultural research in OECD countries was adjusted to requirements of the private sector by means of a number of institutional changes. In some countries (notably in the United Kingdom), public agricultural research was partially privatized, or redirected from near-market research towards basic research, while the industry was often given a greater say in the allocation of public funds (Alston *et al.*, 1998: 69-74).

The second tendency inducing a change in the organization of crop development consisted of an increase in private expenditure on agricultural R&D. There were a number of factors that raised the interest of large industrial capital in plant breeding: economic and political limitations imposed on fuel-based agriculture, expanding Third World demand for food and seed, and advances in genetic engineering. Private agro-industry stepped up its research investment by an annual growth rate of 5.1% in the 1980s, at the same time that public R&D spending on agriculture was slowing down. Private agricultural R&D expenditure in 22 OECD countries increased from US$ 3.9 billion in 1981 to more than US$ 7 billion in 1993, almost equalling total public R&D expenditure on agriculture (Figure 5.13).[6] There are considerable differences among countries, however. In the USA, for example, private sector agricultural R&D expenditure began to surpass public expenditure already in the early 1980s (USDA, 1995). It is quite difficult to assess the proportion of private R&D expenditure devoted to crop development. Assuming a total world seed market of around US$ 28 billion, our estimate of private crop development R&D in the mid-1990s is between the US$ 2.8 and 4.2 billion.[7]

Figure 5.13 Comparison of public and privately performed agricultural R&D in 22 OECD countries (in billions of dollars)

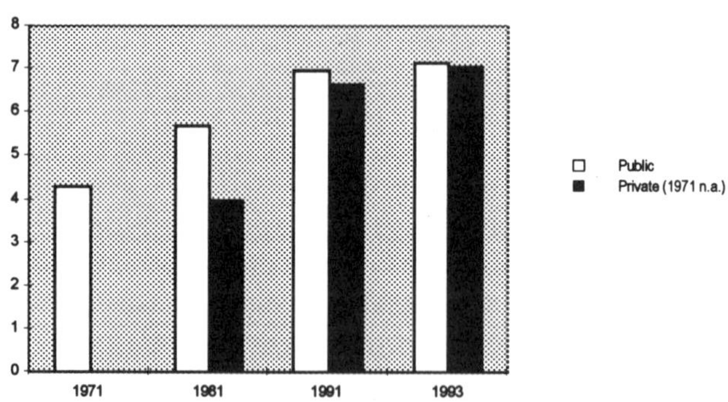

(Data derived from Alston *et al.*, 1998:56,66)

The rise of private investment in agricultural R&D thoroughly restructured the industrial involvement in crop development. Chemical, pharmaceutical, and food industry capital moved into crop development, the existing small-scale plant breeding firms were largely absorbed, and new branches of industry emerged. This reorganization, which is still continuing, has been a condition for the industrialization of crop development. Only by combining the assets of the different industrial sectors, could genetic engineering provide new opportunities for capital accumulation. Companies from at least six different industrial sectors have become involved in crop development.

Plant breeding firms
We have already explained that in earlier Agro-Food Orders the private crop development industry generally consisted of many rural-based firms, often small and family-owned. Although systematic data on the seed industry structure before 1970 are incomplete, the general idea is that the industry was "highly fragmented with extremely low concentration levels" (McMullen, 1987:82). After 1970, the breeding firms were forced to seek financially strong partners, due to increasing competition in the seed market and the development of new, costly plant breeding techniques. In order to survive in the long term, many firms allied themselves to the agro-chemical and food processing industry.[8]

New Biotechnology Firms (NBFs)
The need for linkage with large industrial corporations was also felt by the New Biotechnology Firms (NBFs). NBFs are small, private research organizations, usually university spin-offs, which have been founded to translate new discoveries in genetic engineering into marketable products. The NBFs were basically an American phenomenon, but also emerged in Europe, often upon the joint initiative of governments and large industrial corporations (Orsenigo, 1989:125,152). NBFs that lacked both capital, up-scaling opportunities and a marketing network necessarily relied on venture capital and contracts with client firms, such as chemical and pharmaceutical companies. This trend emerged in the 1980s and continues today.

Genomics firms
In the 1990s, the increasing commercial value of genetic information led to the establishment of NBFs that specialized in the identification and workings of DNA sequences. By using advanced software, genomics firms can scan entire chromosomes for DNA sequences and work on the unravelling of the genetic function of given sequences. This process, generally referred to as 'functional genomics' is targeted at the discovery of genes with potential economic value. The genomics firms operate as contractors for both the larger seed firms and the newly emerging 'life science' divisions of agro-chemical and pharmaceutical corporations.

Software firms
Genome mapping relies heavily on computing power and software for refining the DNA sequencing technologies through advanced software and special DNA- or Gene-Chips. For this reason, software designers have become new, but indispensable actors in crop development. In 1997, for example, Monsanto and IBM announced a strategic alliance to unravel raw DNA data sets by an IBM 'next generation pattern discovery algorithm' (Monsanto, 1998a).

Chemical and pharmaceutical industry
The chemical industry envisaged declining markets for agro-chemicals due to the rising oil prices and an increased environmental awareness. The expected loss of sales could be curbed by producing crops with a built-in crop protection (Bijman, 1999). As molecular biology offers opportunities not only for plant breeding but also for new strategies in drug discovery, the chemical and pharmaceutical enterprises have established 'life science' divisions, in which biotechnology R&D for the two industrial branches have merged. The strategic targets for these large corporations are the 'traditional' plant-breeding firms. They possess the working collections and plant varieties that serve both as source and carrier of plant genetic information. Moreover, the breeding firms have at their disposal a distribution network that can bring the transgenic plant varieties to the farmer. This explains the large sums of money that have been paid to acquire plant-breeding firms. Monsanto, for example, paid a total of US$ 5.3 billion in 1997 and 1998 to take over Holden, DeKalb Genetics and Delta and Pine Land Co. (RAFI press release, 14 May 1998).

Food processing industry
The food processing industry has been investing in crop development because of its interest in adapting the intrinsic qualities of crops, such as nutrient content, perishability and processability. Adaptations in the upstream stages of the production chain would offer better control over downstream food processing.[9] Plant varieties carrying speciality traits also facilitate a trend in the industry towards separating natural products into their basic components (chiefly carbohydrates, proteins and fats) in order to re-assemble them differently (Ruivenkamp, 1989:140-189). Nevertheless, there has been little commercial success as yet with plant varieties that have been genetically engineered specifically for food processing purposes (Duvick, 1996:15-17).

The restructuring of the crop development industry has resulted in the emergence of what can be called *crop development conglomerates*. They comprise close networks of enterprises active in a specific area of genetic crop development. Typically, a crop development conglomerate is organized around one OECD-based transnational enterprise (TNE), rooted in the chemical, pharmaceutical, or food processing industry. This TNE maintains a network of linkages with one or more plant breeding firms, NBFs, genomics and software firms, and also with public research institutes. The nature of the linkages is diverse and varies from temporary research collaboration to complete take-overs.[10] Figures 5.14 to 5.19 present a graphic impression of the nuclei of the six largest crop development conglomerates. For the sake of clarity, the figures show only two types of relationships: acquisitions, and 'alliances' (a broad category that includes all other forms of collaboration). Moreover, the figures include only those participating companies that can be considered the most important partners in terms of their sales. Additional details about the conglomerates can be found in the text boxes that accompany the figures.

Figure 5.14 The Monsanto conglomerate (1998)

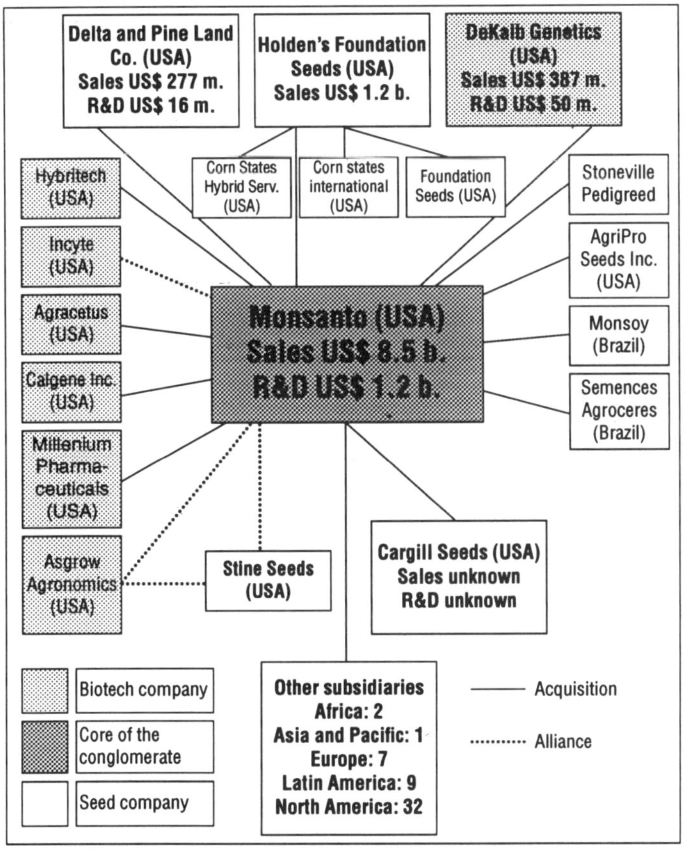

- Between 1996 and 1998, **Monsanto** took over four major U.S. maize breeding companies: **Holden's Foundation Seeds** (price US$ 1 billion) and **DeKalb Genetics Corporation** (US$ 2.3 billion), **Delta & Pine Land** (US$ 1.9 billion), and **Cargill**'s seed divisions (US$ 1.4 billion). Holden and DeKalb respectively cover 35 and 12% of the U.S. maize seed market. The acquisitions made Monsanto the world's largest actor in maize improvement.
- With the acquisition of **Asgrow Agronomics** (US$ 240 million), specialized in maize and soyabean seed, Monsanto has obtained access to about 10% of the U.S. soyabean market. Asgrow, **Stine Seeds** and Monsanto collaborate on genetic engineering of soyabean. **Agripro** is a leading U.S. breeding company, specialized in maize, cotton and wheat.
- With the acquisitions of **Calgene**, a specialist in vegetable seeds, and patent holder of the Flavr Savr-tomato, Monsanto tries to penetrate the vegetable seed market. The acquisitions of the biotech division of **Agracetus**, and **Stoneville Pedigreed** are to increase the involvement in the U.S. cotton seed market.
- The acquisition of **Semences Agroceres** offers Monsanto a 30% market share of the Brazilian maize seed market.
- Monsanto's alliances with **Millenium Pharmaceuticals** and **Incyte** in 1997 will improve the position of its life science division in genomics research.

(Sources: Bijman, 1997:33,35; Cailliez, 1997:42; James, 1997:6; DeKalb, 1998; Monsanto, 1998b; Millennium, 1998)

Figure 5.15 The DuPont/Pioneer conglomerate (1999)

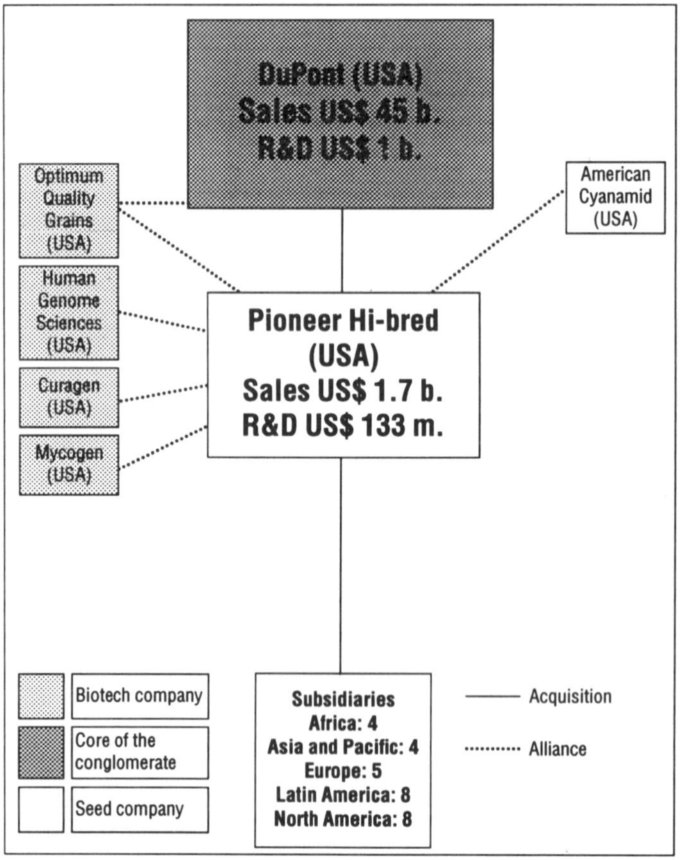

- In March 1999 **DuPont** took over **Pioneer Hi-Bred** for US$ 7.7 billion.
- Pioneer Hi-Bred and DuPont set up a joint venture on a 50/50 basis: **Optimum Quality Grains.**
- In 1997 Pioneer concluded an alliance with **Mycogen** (USA), a new biotechnology firm of which 65% is owned by Dow AgroSciences, the world's no. 7 of the crop protection industry.
- In 1997 Pioneer concluded an alliance with **American Cyanamid**, an American agro-chemical and pharmaceutical firm.
- In 1996, Pioneer established a joint research programme with **Human Genome Sciences** (HGS).
- In 1997, Pioneer invested US$ 7.5 million in Curagen, a genome sequencing firm, and will fund Curagen research for 5 years.

  (Sources: Bijman, 1997:33: James, 1997:6; Pioneer, 1998; Curagen, 1998; DuPont, 1999; Pioneer, 1999)

Figure 5.16 The ELM/Pulsar conglomerate (1998)

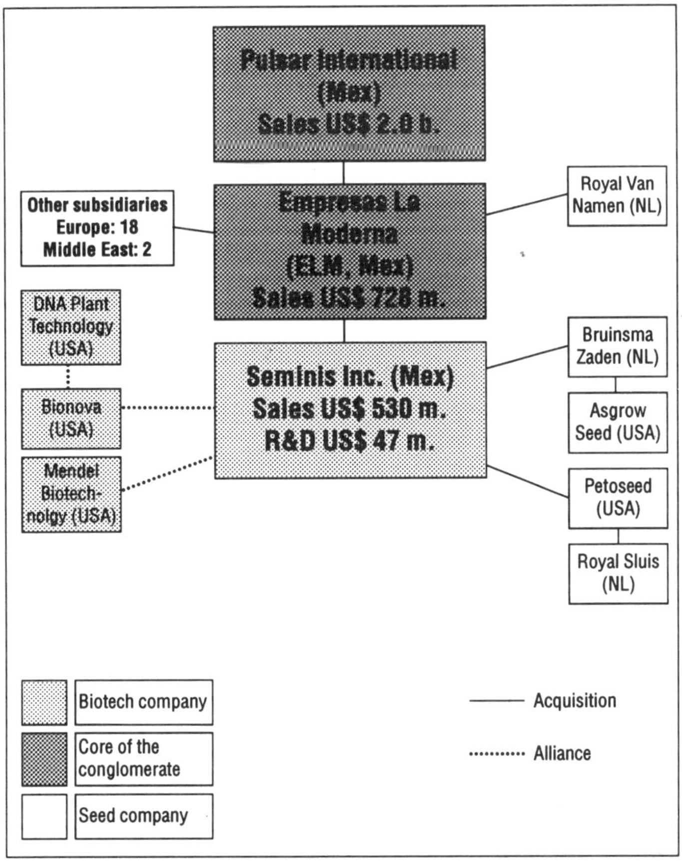

- **ELM** is a subsidiary of **Pulsar International,** a Mexican corporation involved in insurance, financial services and construction.
- In 1994 and 1995, **ELM** invested US$ 480 million in the acquisition of **Asgrow Seed** and **Petoseed** and their respective subsidiaries **Bruinsma** and **Royal Sluis**. All were merged in 1995 to create **Seminis Inc.**
- Due to these acquisitions, **Seminis Inc.** has rapidly become the world's leading vegetable seed company owning 36 breeding stations worldwide and the world's largest vegetable gene bank.
- **DNA Plant Technology** (DNAP) is the world's fourth largest enterprise in agricultural biotechnology.
- Through the acquisition of **Bionova**, a vegetable distribution company, ELM acquired access to the U.S. vegetable market. To gain access to the European and Middle East market, ELM acquired 51% of **Royal van Namen** in 1996.
- In 1997, ELM and Monsanto started a new company **Mendel Biotechnology** (investments US$ 30 m.) for the identification of DNA sequences.

(Sources: Bijman, 1997:33; Cailliez, 1997:42; James, 1997:6; Massieu, 1998:10; Monsanto, 1998b)

Figure 5.17 The Novartis conglomerate (1998)

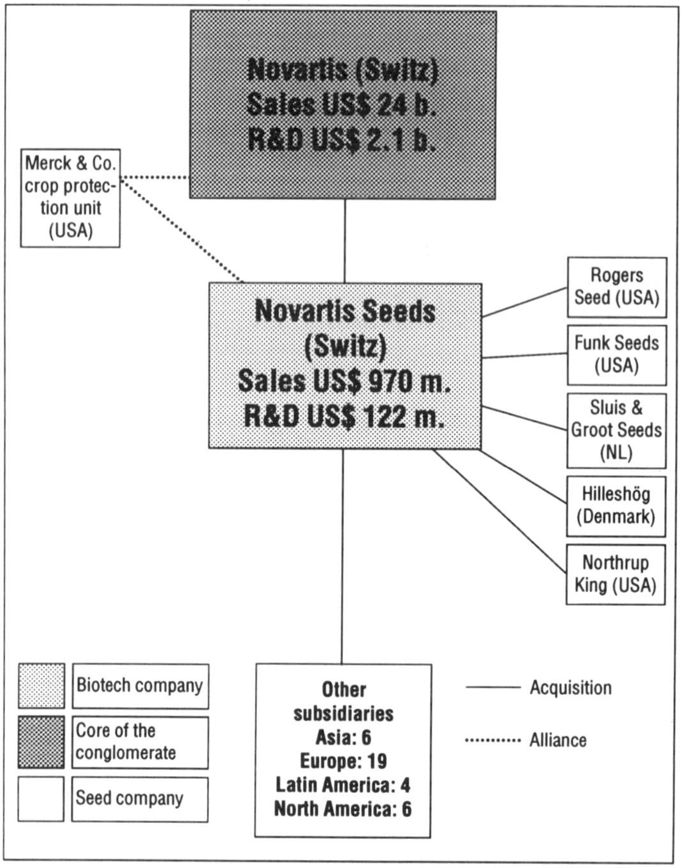

- **Novartis** is the result of a merger between Sandoz (Switzerland) and Ciba Geigy (Switzerland) in March 1996, then the largest corporate merger in history. Novartis holds the number one position in pharmaceuticals (59% of total sales) and in agribusiness (24% of total sales) worldwide.
- **Novartis Seeds** (a subdivision of Novartis Agribusiness) is the second largest seed producer in the world. It acquired:**Northrup King**: specialized in maize, oil seeds and other field crops (worldwide sales); **Hilleshög**: specialized in sugar beets (worldwide sales); **Sluis & Groot Seeds**: specialized in vegetables, sales in Europe, Africa and Asia; **Rogers Seed**: specialized in vegetables, sales in North, Central and South America.
- In 1997 Novartis Seeds acquired the crop protection unit of **Merck & Co.** for US$ 970 m.
(Sources: Cailliez, 1997:42; James, 1997:6; Novartis, 1998a; Novartis, 1998b)

Figure 5.18 The Rhone Poulenc/Limagrain conglomerate (1998)

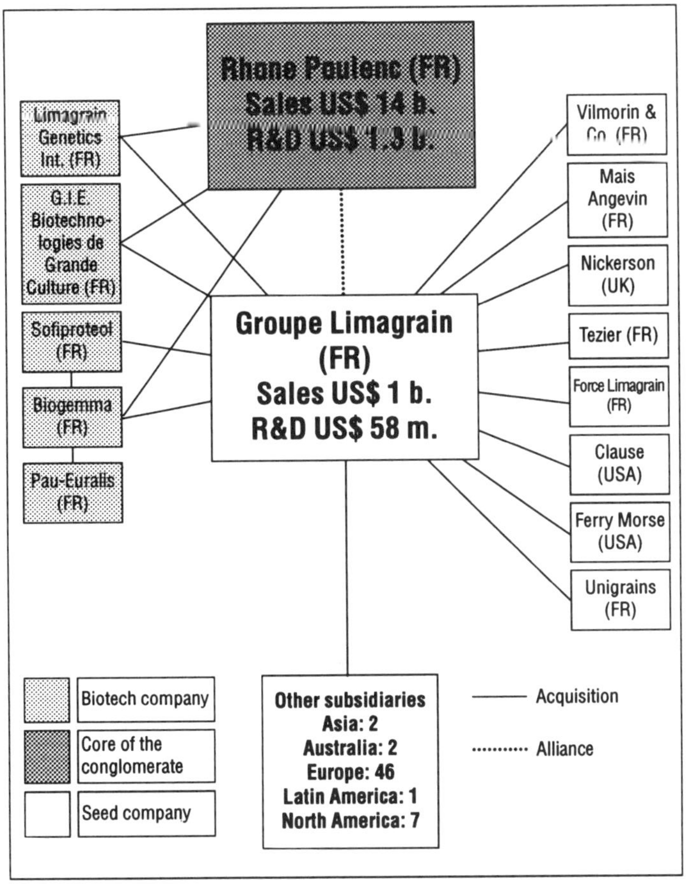

- **Groupe Limagrain** is a cooperation owned by 600 French farmers and seed growers.
- The cooperation is the third largest seed producer in the world and the largest in the European Union.
- Maize seed comprises 25% of total sales; vegetables and flowers 27%; and other field crops 15%. About 33% of Limagrain's sales comprise of processed food and health products.
- In 1994 **Rhone Poulenc Agro** (a sub-division of **Rhone Poulenc**) and **Groupe Limagrain** jointly establish **Limagrain Genetics International** and **G.I.E. Biotechnologies de Grande Culture**.
- Of the 57 European subsidiaries, 31 are located in France, 6 in Spain, 6 in the UK, 5 in Germany, 5 in Italy, 2 in The Netherlands, 1 in Belgium, 1 in Austria, and 1 in Poland.
- France comprises 53% of total sales of Groupe Limagrain, the European Union 29%, and North America 12%.

(Sources: Cailliez, 1997:42; James, 1997:6; Limagrain, 1997; Limagrain, 1999)

Figure 5.19 The Zeneca/Cosun conglomerate (1998)

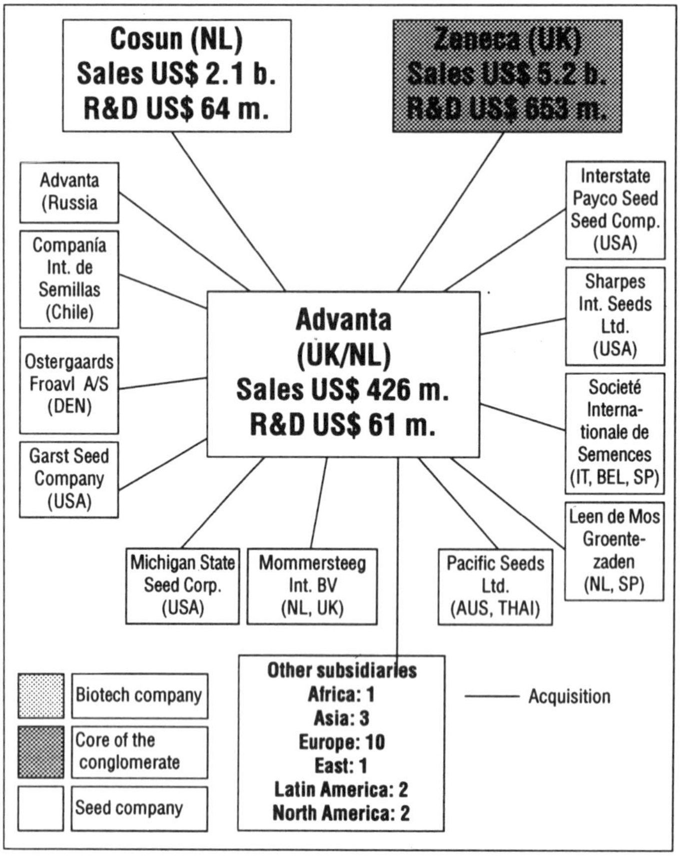

- **Advanta** was created in 1996, and is a joint venture of Zeneca Seeds (a subdivision of **Zeneca**) and VanderHave Seeds (a subdivision of **Cosun**). Zeneca is a large agro-chemical and pharmaceutical company. Cosun specializes in food processing (sugar, fruit, potatoes) and food conservation. Zeneca and Cosun each control 50% of the shares of Advanta. Headquarters are located in The Netherlands.
- Advanta sells seed under the original name of its subsidiaries. Most important Advanta brands are: **Leen de Mos** (vegetables, Europe), **Mommersteeg** (various, worldwide), **Garst** (maize, USA), **VanderHave** (various, worldwide), **Zeneca Seed** (maize, USA). Seeds produced by Advanta also carry the name **Advanta** or **Zeneca.**

(Sources: Advanta, 1997; Cosun, 1997; James, 1997:6; Mooney, 1998:149; Zeneca, 1998)

The emergence of the industrial conglomerates has had profound effects on the organization of crop development. The seed industry seems rapidly to be losing its fragmented nature. Many plant-breeding firms are integrated in one of the conglomerates that are financially and technologically directed by one or two large transnational enterprises. American industry representatives predict that eventually there will be "probably only five companies or alliances in the world that have access to both high-performing germplasm and proprietary biotech research." (Freiberg, 1997c:6).

It would be an overestimation, however, to consider the formation of the conglomerates as a unilinear tendency, inevitably heading towards a monolithic 'life science' oligopoly. There are wide differences between the conglomerates regarding the industrial background of the central corporations, the crop focus, the home-base country, and corporate culture. Moreover, the choices and strategies of the conglomerates, for example as regards the technologies employed, the type of intellectual property sought, and the conservation policy followed, provoke countervailing reactions. These reactions may originate from other private sector sources involved in the production chains, such as food processing corporations or supermarkets, which may strongly influence the demand for specific qualities of new plant varieties. Reactions may also come from the breeding firms that are not (yet) integrated into one of the conglomerates. As we indicate in section 5.3.3, European plant breeding associations in particular differ from genetic engineering companies in their views on how genetic information and plant varieties should be protected as intellectual property. Countervailing reactions may also come from consumers, and a wide variety of civil pressure groups, such as bio-ethical, religious, environmental and farmers' organizations. The responses of all these organizations have to be reckoned with, as the public opposition to plant patenting and the introduction of transgenic plant varieties into Europe shows.

Public reactions to the innovations of the crop development conglomerates reveal not only that there are certain limits to the freedom of industrial initiatives, but also that these limits are no longer set by the government alone, but also by other private, market forces. The industrial conglomerates are obviously the strong dynamic forces, but consumers and public interest organizations can no longer be ignored. The direct interaction of conglomerates with other private industry and consumers is influencing the direction of crop development research. This is what could be called 'market-led' crop development.

The role of the national governments in such market-led crop development is no longer that of the central organizer. We argue that the private crop development conglomerates, rather than the governments, are becoming *the* dynamic force of crop development of the Third Agro-Food Order. This corresponds with the way the industry is depicting itself. In their publications and Internet homepages, the key corporations of the conglomerates tend to equate their commercial intentions with public, common interests. They suggest that they are destined to save the world from food shortage and to preserve the planet by making agriculture sustainable.[11] Notable is the global focus of the enterprises. Food production and consumption is no longer presented as a national issue, as during the Second Agro-Food Order, but as a matter of global concern. As the world population is expected to grow to around 8 billion by 2020, a doubling of the 1997 yields is thought to be necessary (ASSINSEL, 1996). This goal is to be achieved by adapting the main plant varieties to climatic variation, especially with the objective to increase yields on less fertile soils, and by stepping up efforts to increase agricultural productivity.[12] These extra research investments require, according to the conglomerates, new crop development policies that further stimulate private investments in the exploitation of genetic information. New conservation strategies are to secure long-term access to biological material containing commercially interesting genes, while new IPR policies are to safeguard high investments in R&D.

Summary

We have mapped out three broad developments that indicate the emergence of the Third Agro-Food Order. Firstly, the international division of labour in agriculture has not essentially changed in the past decades. The OECD countries as a whole have been able to strengthen their agro-food trading position vis-à-vis developing countries, despite increases in production in the latter. Secondly, at a regulatory level there is a tendency for governments in OECD countries to reduce their presence in the agricultural sector, while private industry, consumer and civil pressure groups become more involved in agro-food production. This development can be considered as a shift in the prevailing agricultural production strategy, away from a basically state-led strategy, towards one that is more market-led. Thirdly, a division of labour also emerges in crop development, with the OECD countries as the main exporters of new, commercial plant varieties. The heavy investment in R&D and the future scenarios of a handful of industrial conglomerates indicate that they are becoming the main organizers in crop development.

5.2 DNA as a new conservation target

Earlier we have described how, during the First Agro-Food Order the interest in plant collection shifted away from individual plants of botanical species to landraces. Especially the settler colonies, notably the USA, were in great need of landraces of temperate crops to develop a competitive agricultural sector. The collection of landraces was intensified in the Second Agro-Food Order, which brought forward even a global conservation system: a close connection of seedbanks established in most countries of the world. Plant breeders extensively used this system and freely exchanged germplasm in order to breed new varieties.

In the Third Agro-Food Order the target and means of plant collection and conservation are subjected to considerable changes. The plant seems to loose its exclusive position as sole resource for the creation of new plant varieties. For genetic engineers within the crop development conglomerates, the pool of genetic resources is much broader than that stored in the seedbanks. Apart from plants, in fact all biological organisms are potentially useful for breeding, as long as they contain the DNA that encodes for a desired trait.

In this section we examine how the crop development conglomerates design new strategies in collecting genetic information, and what impact they have on the existing global conservation system. If private investment in the genomics industry continues to grow at the current pace, and genetic information of micro-organisms, insects, birds, or human blood is screened and stored in 'DNA data banks', will there still be a function for the seedbanks in the future?

Prominent is also the issue regarding the position of the suppliers of genetic information. The new broad genepool covers genetically diverse biological organisms that can be found in tropical forests and in farmers' fields in developing countries. In order to find and identify useful DNA sequences from these organisms, crop development conglomerates have embarked on 'bioprospecting' programmes for the identification and evaluation of genetic properties in virtually any organism. What benefit may developing countries derive from their supplier function?

5.2.1 DNA as an asset in the formation of conglomerates

Advanced gene isolation and transfer technologies have allowed the crop development industry to introduce a wide range of genetic material into a host organism. Most engineered crops that now appear on the market, notably in the USA, contain a single added gene that confers a single desired characteristic, such as resistance to specific insect or viral pests, longer shelf-life, or tolerance to a specific type of herbicide. Monsanto, DuPont, and Pioneer Hi-Bred can be considered as forerunners in this field. Most of the genetically engineered or modified crops they produce contain 'foreign' genes, for example genes extracted from bacteria.

As indicated above, the commercial interest in finding commercially attractive genes has resulted in the emergence of a new, very capital-intensive branch of the crop development industry, the 'genomics' firms. They operate as contractors for the core corporations in the conglomerates, which allocate a great deal of their investment to DNA sequencing. Pioneer Hi-Bred, for example, has signed a contract with one of the largest American genomics firms, CuraGen, "to mine the Company's extensive germplasm base to identify particular genes that promote superior traits in plants." (Pioneer Hi-Bred, 1998a:16). Pioneer is also involved in a US$ 40 million maize genomics programme with DuPont (RAFI, 1998a:4). In 1997 Monsanto concluded a US$ 218 million licence agreement with Millennium Pharmaceuticals (USA) and IBM to exploit genome data. Monsanto plans to build a new "genomics research centre", staffing 100 scientists, located next to Millennium's headquarters (Osborne, 1997:1). The agreement with Millennium will cover "all plants in all countries". Monsanto considers the agreement to be an integral part of its life sciences strategy, and hopes to "gain a competitive edge in the search for patentable crop genes" (Monsanto quoted in RAFI, 1998a:4). Novartis, in 1998, announced that it would invest the record amount of US$ 600 million to establish the Novartis Agricultural Discovery Institute, which is entirely dedicated to genomics studies, especially in respect of vegetables (*AgBiotech Reporter*, August 1998:1).

The new genomics industry is closely associated with producers of advanced software and hardware (DNA-chips) that help to speed up the analysis of the workings of unique DNA sequences. The doubling of the computer chip's power every 18 months rapidly increases the accuracy and scope of DNA-markers and reduces screening costs per unit of genetic information. In the words of the DuPont president for France, Francis Holliday: "Three years ago it would take a DuPont scientist two years and a half million dollars to identify one gene. Now I call up my scientists and say how many [genes] did you find on Monday, and they will say 600 at a cost of US$ 100 per gene. The velocity of scientific research is astronomical and I don't see it stopping." (Francis, 1998)

The prospect that national competitiveness in agro-food production will largely come to rely on genetic engineering, molecular breeding and related genomics research, has instigated a redirection of government investment in agricultural research. Major public projects have been initiated since the mid-1990s, in which millions of DNA base pairs on the chromosomes of key crops are determined and analysed. For example, the U.S. National Centre for Genome Resources (NCGR) specializes in "data warehousing for gene sequences", supplying DNA sequence data to both public institutes and private firms in the medical and agricultural sectors. The NCGR is rooted in the Human Genome Project and works free of charge (NCGR, 1998). In the USA, genome projects currently focus on more than 50 different crops (USDA, 1999). Europe and Japan have set up genome projects on 16 and 24 crops and organisms respectively (MAFF, 1999).

How close the genome projects are linked to agro-food industry interests in OECD countries is illustrated in the example of the U.S. Maize Genome Project. In 1997, the Missouri Corn Growers Association and the U.S. National Corn Growers Association (NCGA) started to lobby for a US$ 143 million National Corn Genome Initiative. The NCGA claimed that a DNA sequence map of the entire maize genome would contribute to the identification of major genes that would boost harvests by an average annual increase of 3% (equivalent to US$ 1 billion extra revenue per year) (NCGI, 1999). Sequencing the genome of maize would consolidate the U.S. dominance in the world maize market. It also would increase the diversification of maize-based products (already embracing 3500 items) and decrease the dependence on chemical inputs due to higher resistance (NCGI, 1998). The Corn Genome Initiative is to form the core of a US$ 200 million 4-year National Food Genome Project which, in turn, is the cornerstone of a US$ 780 million five-year Initiative for Future Agriculture and Food Systems (Anonymous, 1997).

Because computer-aided analysis of genetic sequences has become essential for the creation of new plant varieties, DNA data banks have emerged as important reservoirs of genetic information. The current practice of putting the output of public genome projects on freely accessible Internet sites may therefore soon be endangered. Already, genome projects financed by Japan, China and various European governments are being criticized by U.S. researchers for their limited release of sequencing data. Similar accusations have been made against Monsanto and DuPont (Sommerville, 1997). Because most public DNA data banks are explicitly created to increase the competitive edge of a specific commercial crop (predominantly maize and soyabean), the practice of shielding 'raw' DNA data from competing states or conglomerates is likely to become more common. Protective measures may appear in genome projects in which private and public interests strongly overlap, as in the case of the U.S. National Corn Genome Initiative.

The broadening scope on collecting genetic information has also provided an extra incentive for mergers and acquisitions of plant breeding firms with economically interesting 'working collections'. The recent series of acquisitions and mergers by Monsanto were strongly related to the company's strategy to gain access to the 'gene libraries' of the breeding firms. Regarding its acquisition of Biopol, Monsanto explicitly stated that Biopol "provides the genes and associated technologies to develop biodegradable plastic polymers by fermentation and in plants." (Monsanto, 1998b).

High investment in DNA sequencing indicates that the seed based conservation strategy of the Second Agro-Food Order is rapidly changing. The budget of Novartis' genomics project alone, around US$ 600 million, is twice the total annual budget of the CGIAR. Conservation in the form of DNA sequences in combination with computerized monitoring systems has, for the first time, allowed the crop development industry to screen vast genepools for useful genetic information.

5.2.2 Bioprospecting for DNA data

Genetic engineers can tap from a much larger pool of genetic information than 'classical' breeders. Whereas classical breeders concentrate on genetic information in landraces and varieties which are genetically familiar, a genetic engineer seeks to exploit genetic information from any organism. New developments in genetic engineering and genomic research have created practically infinite genepools. The attitude of genetic engineers in exploiting these genepools is as straightforward as it is ambitious. It is assumed that for any desired trait "there will be an organism that is able to do the job". The aim is to find that organism, identify the genes involved, and insert them into a suitable crop (Nap, 1999).

The broadening genepool has induced the life science divisions of the core corporations in the conglomerates to embark on 'bioprospecting' programmes in which virtually every organism may be screened for unique DNA sequences. Although any other form of 'biodiversity'[13] may be useful as long its contains genes that encode for a desired trait, genetic engineering has put in the spotlight a category of plants that was always considered least attractive to breeders and genebank curators: the wild or 'undomesticated' ancestors of crop plants (*cf.*

Tanksley and McCouch, 1997:1063; Rejesus *et al.*, 1996).

The interest in biodiversity as a future source of genetic information has been spearheaded by large pharmaceutical companies, such as Merck & Co. and Glaxo Wellcome. Pharmaceutical corporations took the initiative in setting up bio-prospecting programmes in the early 1990s, particularly in the tropical forests of developing countries (Reid *et al.*, 1993:8-9). The messages on the internet sites of Monsanto and Novartis reveal also that enterprises with an agro-chemical background are increasingly interested in biodiversity as a source of genetic information for crop development (Monsanto, 1998b; Novartis, 1998). The Novartis Agricultural Discovery Institute, currently the world's largest private genome project, has announced a close collaboration with Missouri Botanical Gardens (USA) to secure its access to exotic species with potentially useful DNA (*AgBiotech Reporter*, August 1998:1).

Industrial lobbying organizations, such as the 'World Business Council for Sustainable Development' (WBCSD) have taken an interest in the industrial application of biodiversity. The WBCSD was set up to provide a business perspective on sustainable development for the UNCED. With a membership of more than 120 (often transnational) companies from 36 countries, the council is the most important spokesman for business interests in the environmental debate (WBCSD, 1995). The WBCSD has prepared a joint publication with the World Conservation Union (IUCN) under the title "Business and Biodiversity", in which it is indicated that wild relatives of major crops and other forms of biodiversity may well become the crucial link between environmental conservation programmes and future crop development strategies (WBSCD and IUCN, 1997:12).

The collaboration between industry and international environmental organizations is effective as the latter have a strong impact on the implementation of the CBD. Since 1992 they have participated in the Global Biodiversity Forum (GBF), the secretary of which is based at the IUCN headquarters in Gland, Switzerland. The GBF is a spin-off of the Global Biodiversity Strategy that was proposed by the World Resources Institute (WRI, Washington D.C.), the United Nations Environment Programme (UNEP, Nairobi), and the IUCN. The GBF describes itself as a "mechanism to foster analysis and unencumbered dialogue and debate among those interested in priority ecological, economic, institutional and social issues" related to biodiversity. The GBF acts as an umbrella organization for all actors interested in the conservation of biodiversity, including NGOs, governments and private industry. Most funding is provided by WRI, IUCN, and in particular the Global Environment Facility (GEF) (Ruff and Worthington, 1998:8).

The long-term interest of OECD countries in biodiversity is reflected in the new and relatively large funds generated by the GEF. This facility was created just before the UNCED in 1992 - the conference that also produced the Convention on Biological Diversity (CBD). Although the United Nations Development Programme (UNDP), UNEP and the World Bank jointly administer the GEF, the World Bank reportedly has a decisive say in its funding policy. Since the UNCED, the GEF has provided for large funds to allow CBD signatory countries to finance biodiversity conservation programmes. In the period 1991-1994, the GEF reserved US$ 560 million of its total budget of US$ 1.3 billion for the conservation of biodiversity (GRAIN, 1998).

GEF funds have also allowed the CGIAR, another organization under the World Bank umbrella, to increase its investment in biodiversity. In the period 1991-1994 "environment" and "biodiversity" related projects totalled about US$ 200 million, about 20% of the CGIAR's total budget in that period (CGIAR, 1997:42). This expansion of biodiversity conservation enabled by the GEF contrasts with the stagnant budget allocated by CGIAR for the maintenance of its seed collections in the same period, estimated at US$ 100 million (Strauss, 1995:2). The new budget for biodiversity allowed IPGRI (previously IBPGR, the CGIAR institute responsible for conservation strategies) to increase significantly its focus on biodiversity. In 1993, the CGIAR established the System-wide Genetic Resources Programme (SGRP), covering crops and their wild relatives, forage, forestry, and agroforestry species, livestock and aquatic genetic resources with the explicit aim to "... enhance the efficiency, effectiveness and transparency of CGIAR's contribution to the Convention on Biological Diversity ..." (IPGRI, 1998). IPGRI has started to hire out its expertise to CBD signatory countries implementing new biodiversity projects with GEF funds.

The biodiversity-oriented conservation strategy of IPGRI is notable for an organization that has specialized in the past 25 years in collecting landraces for the main food crops. Like many other seedbanks that are under budgetary pressure, the broadening of IPGRI's conservation strategy should be related to its institutional interest to secure its share in the GEF funds for biodiversity conservation projects. The rapid development of "agro-biodiversity" projects since UNCED by IPGRI and other CGIAR institutes indicate how the (funding) gap between biodiversity and landrace conservation is being narrowed.

In short, biodiversity conservation is becoming 'big business'. This is not so much because the public environmental concern in OECD countries has increased, but because the industrial interest in DNA sequences of exotic species has turned the tropical forests into new storehouses of genetic information. The substantial GEF funds for biodiversity conservation indicate that governments of OECD countries are well aware of the economic value biodiversity represents. As a result, landrace seed conservation, the main activity of most seedbanks during the Second Agro-Food Order, has been downgraded to just another branch of biodiversity conservation.

5.2.3 Seedbanks as secondary sources of genetic information

The investments in DNA data banks and genomics research signal the growing dominance of the genetic engineering paradigm over classical plant breeding in crop development. What are the consequences of the new paradigm for the traditional bond between plant breeding and landraces? While they have traditionally been the primary conservation target of plant breeders, in the Third Agro-Food Order landraces have become just one resource of genetic information among many other biologically diverse organisms. It seems therefore inevitable that the seedbanks as traditional storehouses of landraces are losing their key position in the conservation policies of OECD countries.

The declining importance of landraces for crop development is reflected in the shrinking governmental funding for public seedbanks. National seedbanks in many OECD and developing countries face reductions of their budgets (Anonymous, 1992:13). In 1998, the USDA froze its contributions to the U.S. National Plant Germplasm System (NPGS). The U.S. government has also gradually decreased its contribution to the total US$ 300 million CGIAR budget from 28% in 1986 to 12 cent in 1996 (Pardey *et al.*, 1996:16; Powledge, 1994:4). In spite of increased contributions from European countries, the tighter financial situation has caused CGIAR seedbanks, notably those of IRRI and CIMMYT to cut down their activities in seed conservation (Duvick and Shands, 1994:53).

Now that genetic information can be tapped from a much broader genepool than before, DNA data banks will affect the position of seedbanks as primary stocks of genetic information. Private investment in genomics research and DNA sequence analysis and storage is many times higher than the public spending on seed collection. Decreased state support has forced the NPGS to lobby for funding within industrial circles (press release of the American Genetic Resource Alliance, 1998). So far the campaign has only resulted in the American Seed Trade Association calling for a doubling of the American public budget for seedbanks from US$ 20 to 40 million (ASTA, 1998).

The modest corporate interest in seedbanks is not surprising. From a commercial point of view, the collection and storage of genetic information in the form of data on DNA sequences is much more attractive than in the form of seed. DNA sequence technology in fact frees the crop development industry from a problem that is as old as professional breeding, namely incomplete information on the genetic content of the crossing material. The associated risk of 'genetic pollution' of the parental line is quite large when landraces are used as crossing material. The 'art' of breeding is to single out the desired characteristics from the broader genetic background of a landrace. This laborious and time-consuming work was usually carried out by the public breeding sector. The resulting elite lines were used as a basis for further breeding in the private sector. However, the more public plant breeding is subjected to market conditions, the more landraces are avoided. Currently, both public and private breeders tend to use those 'improved' plant lines and varieties in which the most important characteristics have been concentrated and from which most undesirable traits have been eliminated. Only when this existing material cannot give a clue to a required varietal characteristic, will plant breeders resort to landrace seed in the seedbanks.[14]

The new interest in DNA sequences also has implications for collection strategies. It appears that there is a decline not only in the crop development industry's interest in seedbanks as resources of genetic information, but also in the public funding for seed conservation. The main reason is that the competitive strength of the agricultural sectors of OECD countries does not lie in servicing the crop development industry with seedbank collections alone, but in the exploitation of genetic information in a much broader genepool at the DNA level. Searches for landraces as they took place throughout the First and Second Agro-Food Orders will therefore be embedded in much broader collecting initiatives, such as bioprospecting programmes in which landraces constitute only a fraction of the species targeted.

5.2.4 A re-nationalization of conservation?

The growing interest of OECD countries in biodiversity runs parallel to the growing awareness in developing countries that the great variety of biological material in their territory has a potential economic value. To take advantage of their national biological resources, developing countries started to question the *de facto* 'common heritage' status of seed, plants and plant material in the late 1980s. The status allowed researchers from OECD countries to collect seed and plant material freely, all over the world. During the UNCED in 1992, developing countries managed to replace the common heritage principle by the principle of national sovereignty.

The CBD in which the principle of national sovereignty over biological diversity was affirmed, was signed and ratified by all 174 countries except for the USA. The position of the USA is somewhat peculiar. Its reason for not signing is reportedly related to the position of the Republicans, who expect developing countries to milk the CBD for cash. The crop development conglomerates and the U.S. Biotechnology Industry Organization, however, are in favour of ratification. They argue that not ratifying the CBD might affect their interests in the export of genetically modified organisms (GMOs) when the CBD is extended with a Biosafety Protocol, early in 1999. The industry is concerned that other CBD member countries may use the non-party status of the USA to impose restrictions on the import of U.S. produced GMOs (Anonymous, 1998).

Indeed, the CBD does strengthen the position of developing countries in their role as exporters of biological material with potentially useful genetic information. The CBD provides that the collection of biological material requires "prior informed consent" from the "country of origin". The purpose of the collection mission and the further use of the material collected must also be explained. The CBD stipulates that the supply of biological material shall be compensated. Since 1992, almost all CBD signatory countries have been reconsidering their conservation policies. Nevertheless, only a limited number of developing countries, including Brazil, Chile, Colombia, the Philippines and Kenya, have established and implemented a new regulation for the export of biological material. Most countries do not have the infrastructure to implement or even design such regulations.

At first glance, the implementation of the CBD seems to have brought an end to the global conservation system that formed an important pillar of the crop development industry during the Second Agro-Food Order. The CDB eliminated the 'commons' status of biological material, and defined all biological material within a country's borders as national patrimony. In other words, the CBD implies a re-nationalization of plant conservation, allowing developing countries to demand compensation for the export of biodiversity with economically valuable genetic information. Nevertheless, it is still questionable who is to benefit most from the CBD. There are a number of indications that the new international conservation regulation developing in the Third Agro-Food Order is still primarily global in scope, and particularly directed at the interests of the OECD-based crop development conglomerates.

Firstly, the CGIAR has succeeded in sustaining the 'old' global conservation system, based on the principle of common heritage, for seeds of landraces collected before 1992. To this end, the CGIAR concluded an agreement with its previous rival, the FAO Commission on Plant Genetic Resources in 1994. The agreement brought all CGIAR *ex situ* collections under the auspices of the FAO Global System on Plant Genetic Resources for Food and Agriculture, which is governed by the FAO International Undertaking on Plant Genetic Resources. This remarkable agreement indicates CGIAR's strong interest in keeping 'its' collections out of the newly emerging CBD conservation system based on country specific 'material transfer agreements' and payments. Maintaining its collections in the public domain is a primary condition to sustain the free flow of landraces. Indeed, subjecting the access and use of landraces to the national sovereignty of the countries of origin could open a Pandora's box for the CGIAR, and raise practical, bureaucratic, and financial obstacles (Pistorius, 1995b:20).

Secondly, the economic value of biodiversity is almost entirely conceptualized by a new and influential school of 'economic botanists' in the UK and USA. The main line of thinking of this school is to organize the conservation of biodiversity on the basis of economic incentives. Key advocates are the World Resources Institute (WRI), the IUCN, and economic faculties at the Stanford, Harvard and Oxford Universities. They advise the OECD, GEF and FAO with publications such as: *The Appropriation of the Benefits of Plant Genetic Resources for Agriculture* (Swanson *et al.*, 1994), The *Economic Value of Biodiversity* (Pearce and Moran, 1994) or *Genes for Sale: Privatization as a Conservation Policy* (Vogel, 1994). The bioprospecting programmes spearheaded by the WRI in the early 1990s form the most notable exponent of this school of thought. The result is a utilitarian perception of biodiversity that encourages developing countries to invest in the conservation of biodiversity, while the difference between the actual and potential value of biodiversity for the biotechnology industry often remains unclear.

Thirdly, given the fact that the global market for 'raw' genetic information is highly specific and therefore 'demand-driven', the developing countries, by signing the CBD, will depend on marketing conditions set by crop development conglomerates. If a new 'market' for biodiversity emerges, its potential economic value can only be assessed and determined by these conglomerates.

Fourthly, although developing countries may improve their services as suppliers of diverse genetic information and may become well-equipped biodiversity treasurers, they remain providers of raw material. The industrial exploitation remains in the hands of the foreign conglomerates, because they have the financial and technological capacity to exploit genetic information. The bioprospecting programmes that have been initiated so far (like that of the pharmaceutical corporation Merck and Co. (USA) together with the *Instituto Nacional de Biodiversidad* (INBio) in Costa Rica), show no long-term prospect of developing country organizations appropriating more value-added stages in the industrial exploitation of their national biodiversity (Pistorius and Van Wijk, 1993).

Fifthly, it is unlikely that developing countries are going to earn much from supplying raw biological material to foreign industrial conglomerates. So far, there have hardly been any occasions in which bioprospecting agreements have generated real financial benefit or other profit for the country of origin. From the information that has been disclosed about biodiversity screening contracts, it appears that the royalties do not exceed six%, and are usually set at about 2-3% in the rare case that the genetic information is eventually processed into a marketable product (Pistorius and Van Wijk, 1993).

Finally, even though developing countries have national sovereignty over their biological resources, their conservation relies heavily on funding from OECD countries. Governments from these countries thus have the opportunity to set the overall terms of conservation programmes, including the conditions under which the collected material is made available to funding countries. Private funds, from OECD-based pharmaceutical and agro-chemical corporations in search of unknown chemical substances and genetic information, are also becoming prominent in the finance of conservation projects in developing countries. As mentioned in section 5.2.2, the future corporate interests in biodiversity are already explicitly recognized by the WBCSD.

5.2.5 A new global conservation regime

The use and storage of genetic information in the form of data on DNA sequences seems to mark the beginning of a new era in conservation policies. Accelerating investments in genetic engineering are likely to enlarge the gap between developing and OECD countries in the international division of labour in crop development. As the Third Agro-Food Order has just started to unfold, it remains a difficult task to assess the long-term implications of the concentration and privatization of crop development for global conservation strategies. Nevertheless, we would like to make an attempt to elaborate further on the trends depicted in this section.

The importance of landraces and improved varieties as primary sources of genetic information for the crop development industry is declining. Crop development conglomerates are spearheading a development in which the genepool is covering practically all diverse biological organisms. Landraces stored in seedbanks constitute only one element in this ‘ocean’ of genetic information which are potentially useful. In an effort to adjust to the new breeding requirements, seedbanks may attempt to add value to their collections by subjecting them to 'in house' DNA analysis. Others may embark on programmes to supply non-industrialized, organic, farming systems with landraces. The limited public funds available for such programmes, and the lack of interest from the private sector suggest however that many seedbanks face a difficult future.

Although seedbanks are no longer the key suppliers of genetic information in OECD countries, they remain important for developing countries. As most crop development in the Third World does not take place at the molecular level, the interest in access to national and international CGIAR seedbanks is likely to remain stable. The publicly accessible DNA data banks of OECD countries (notably Japan, the UK, and the USA) are probably not useful for institutions and companies that do not operate in the forefront of genetics. Only very few developing countries (such as Brazil and India) seem to have sufficiently advanced genetics programmes to use DNA data in crop development.

In spite of the technological and economic barriers that prevent developing countries from exploiting genetic information in the form of DNA sequences, they will remain the main suppliers of that information. Most diverse genetic information can still be found in the wild, particularly in the tropical forests of developing countries, the 'epicentres' of biodiversity. OECD countries thus have a long-term interest in facilitating developing countries in the protection of their biodiversity. OECD-based international environmental organizations, supplied with GEF funds, will further help governments of developing countries to carry out their new responsibility as protectors of their national biological heritage. The CBD, in turn, creates the international regulatory framework to govern the exchange and access to genetic information on the basis of national sovereignty.

While the emerging biodiversity-oriented conservation regime fosters developing countries to act as custodians of the biodiversity within their territory, the ultimate benefit for the national agricultural sector seems limited. Farmers and breeders using classical crossbreeding techniques rely on landraces or existing varieties, but have little direct interest in DNA sequences derived from a broad range of other species. There is thus only a limited connection between national crop development and new, often GEF funded, biodiversity conservation strategies in developing countries (Wood and Pistorius, 1993:18).

The growing disparity between national crop development and global conservation raises questions about the future position of the CGIAR. At the moment, its strategy seems somewhat ambivalent. As the new IPGRI policy indicates, the CGIAR may well become a new key player in categorizing and collecting wild relatives of crop plants or other organisms. Whether the CGIAR intends to use this information for advanced agricultural research in developing countries, remains unclear. Simultaneously, some CGIAR institutes such as CIMMYT and IRRI have embarked on programmes that supply small-scale local farming systems with more genetically diverse seed. The CGIAR's future conservation strategy will probably depend on a decision whether to support non-industrialized or industrialized farming systems.

Summary

Seed of landraces was the primary target for conservation during the Second Agro-Food Order because of the genetic 'familiarity' with the parental lines of the most important food crops. However, since genetic engineering allows for the elimination of species barriers, genetic similarity is no longer a prerequisite for crop development. Only the crop development conglomerates and a limited number of public institutions in OECD countries have mastered this crop development technique.

The new DNA oriented crop development industry has also induced searches for specific sequences in very large genepools. Landraces have been downgraded to 'just another' resource of genetic information, not more or less valuable for crop development than any other organism. The decline of the importance of landraces for crop development is affecting the position of seedbanks as suppliers of genetic information in OECD countries. Their role is overshadowed by bioprospecting programmes that aim at extracting genetic information from a far broader genepool than the landrace collections in seedbanks represent.

The fact that most unique DNA sequences can be found in the biodiverse forests of developing countries has renewed their role as supplier of genetic information. However, the direct utility of DNA sequences for crop development in developing countries is minimal.

5.3 Patents protect plant-related inventions

The reorganization of crop development that took off in the 1970s was not only related to changes in plant conservation policies. The increasing prominence of industrial conglomerates in crop improvement was facilitated by - and required considerable changes in - intellectual property right (IPR) protection. Crop development policies during the Second Agro-Food Order included legal plant protection against the unauthorized propagation of plant varieties. In order to protect the interests of plant breeding firms, governments of various industrialized countries established plant variety protection (PVP), and internationally harmonized their legislation with the founding of UPOV in the 1960s. The UPOV system was still in the process of implementation in the 1970s, when the restructuring of the crop development sector set in, and a profound change in the industrial requirements for such protection was already heralded.

Several developments stimulated the industrial interest in IPR protection in crop development. In the first place, private investment in crop development expanded considerably. As we pointed out in section 5.1.2, private investment in agricultural R&D increased annually by more than five% in the 1980s, and in the mid-1990s private and public investment in agricultural R&D was almost equal. Therefore, there is more economic value to protect. Secondly, the number of competitors in crop development increased. The involvement of chemical and pharmaceutical corporations with plant breeding, and the foundation of NBFs, revealed that private plant breeding was no longer confined to rural, often small-scale family firms. Thirdly, the formation of crop development conglomerates, and the many ensuing collaborative linkages, required precise demarcation of property. IPRs had to be assigned over mutual contributions and joint research results. Finally, with the increasing global scope of the crop development conglomerates, the issue of legal plant protection transcended the borders of OECD countries. Conglomerates export seeds and propagating material and invest in the adaptation of their varieties to the various ecological environments in Asia, Latin America and Africa. In order to play their global role, they demand worldwide protection of their technology and genetic information.

Legal protection of plants has thus become a prominent issue in crop development during the Third Agro-Food Order. However, the reorganization of crop development did not just require more of the existing type of protection. Especially the newcomers, such as the chemical and pharmaceutical corporations, and the NBFs, who had become involved in plant breeding because of their capacity in genetic engineering, showed a preference for (utility) patents[15] for their innovations. This section examines the major changes in IPR protection of plant genetic material during the Third Agro-Food Order. Subsequently we discuss: (a) the revitalization of the debate on plant patenting in the 1980s, (b) the differences in IPR strategies within the crop development industry, and (c) the export of the new prevailing IPR protection regulation to developing countries. We end this chapter with a discussion on the question of how the new IPR protection rules support the international division of labour in crop development in the Third Agro-Food Order, and what effect the new IPR regulation has on plant breeders and farmers in developing countries.

5.3.1 Revival of the plant patent debate

Until the 1980s, private breeding firms were primarily concerned with the 'dual-use value' of their plant varieties. Varieties were to be used to grow a crop, not as propagating material. Unauthorized propagation of the variety could significantly limit returns on investment. In order to curb the dual use of plant varieties two strategies were followed: the release of hybrid varieties and the adoption of PVP legislation. For some private breeders the protection offered was too limited. As we discussed in the previous chapter, the international plant breeders association, ASSINSEL, had actually opted for patent coverage of plant varieties in the 1950s. Since neither the industry nor the European governments allowed the breeders to resort to the patent system, the association had to accept PVP under UPOV as a second best solution. PVP was limited as it did not cover the use of the plant variety's genetic information for further breeding. This had the advantage that all breeders could freely rely on the pool of varieties available - including those of competitors. New varieties were created by crossing or selecting the best performing ones. Also, the American Plant Patent Act of 1930 hardly prevented the use of existing varieties for further breeding. The scope of protection provided by the PPA was so narrow that every close substitute which was not an exact copy of a protected variety (such as an individual naturally occurring mutant) was legal and eligible for protection itself (Stallman and Schmid, 1987:433).

The issue of the open access to the common pool of commercial varieties became prominent when private investment in crop development rose and the industrial conglomerates emerged. PVP did not protect genetic information incorporated into a plant variety from being used by competing breeders, while genetic information could only be protected when incorporated in specific plant varieties. Protection of genetic information *per se* was not possible. For this reason, those companies whose prime concern was the research and modification of plant genetic information rather than the breeding of whole plant varieties, favoured patent protection for their innovations.

The revival in the 1980s of the debate on plant patenting, which had been laid to rest in the 1960s, can be considered as a new phase in the long-term tendency to restrict the triple-use value of plant varieties. In the 1950s, ASSINSEL had achieved a limitation of only the unauthorized commercial propagation of varieties. In the Third Agro-Food Order, however, when large industrial capital has become the central organizer of crop development, the pressure to adjust the subject and scope of IPR protection of plant-related innovations has increased. In addition to propagation, it is the use of a variety's genetic information to create a new variety for which protection is being sought. From an industrial perspective, plant varieties are to be used for growing a crop. All other unauthorized uses of varieties lower returns on investment and must therefore be eliminated.

Whilst it is a continuation of an old controversy, the debate on patenting plants in the 1980 and 1990s is taking place under new conditions. The actors involved in the plant patent controversy are either new, or they have taken a new position. The major proponent of patent coverage for plants is the chemical and pharmaceutical industry - which had resisted plant patenting during the Second Agro-Food Order - and the specialized biotechnology firms. They have an obvious interest in IPR protection of genetic information. Moreover, patent protection is not frustrated by the lack of identification techniques, as it was in the 1950s. The industry is supported by governments in OECD countries. Since the 1980s, the patent offices in most OECD countries have granted patents on plants or plant genetic information. Somewhat remarkable is the change in position of ASSINSEL, which once promoted plant patenting, but which now (as we explain below) opposes patent coverage for plant varieties. Also new in the plant patent debate of the Third Agro-Food Order is the involvement of many new organizations that oppose the grant of patents on plant material and other biological organisms on the grounds of ethical values, farmers' interests, or environmental considerations.

5.3.2 Patent strategy of genetic engineering companies

With the emergence of the crop development conglomerates a split in the IPR strategy of the industry became visible. On the one hand, the genetic engineering companies in the conglomerates are determined to obtain and enforce patent protection for plant genetic information with the aim to eliminate unauthorized use of this information for breeding and propagation. On the other hand, the plant breeding firms (particularly European ones) foster PVP that reduces unauthorized propagation of plant varieties but leaves their use for breeding, in principle, free. In this section we discuss the strategy of the genetic engineers. The strategy of the plant breeders is at issue in the next section.

In order to have plant-related inventions patented, the genetic engineering corporations had to remove a number of legal barriers that were raised in the past to prevent the granting of patents on living organisms (USA) or plant varieties (Europe).

In the USA, the landmark event in this respect was the 1980 decision of the American Supreme Court to change the traditional interpretation of the U.S. Patent and Trademark Office (USPTO) that living products were "products of nature" and therefore not subject matter under the American patent laws. In its Diamond v. Chakrabarty decision on a patent for genetically engineered bacteria capable of cleaning oil spills, the Court held that an invention is not precluded from patent protection because it consists of living matter. It was maintained that the criterion for patentability in this field is whether the invention is the result of human intervention. This Supreme Court decision provided a judicial framework for subsequent USPTO decisions to issue patents for plants and animals (Bent *et al.*, 1987:535). The first U.S. (utility) patent for a plant was awarded in 1985 through a decision of the USPTO Board of Patent Appeals and Interferences concerning the Hibberd patent. The patent involved genetically engineered maize plants containing an increased level of tryptophan, an amino acid.

The new patent opportunities in the USA have considerably expanded the scope of protection for genetic engineers. Rather than visible differences between plant varieties, it is the supposedly inventive technological method for creating the plant, or a specific fragment of its genetic make-up that constitute the basis for protection. A patent on a plant often protects (a) the inventive technique for producing the novel plant 'genotypes', (b) the DNA sequence which is responsible for that, (c) whole plants which express the genetic characteristic, and (d) all plant tissue and progeny of these plants. With this broad scope of protection, the triple-use value of plants can effectively be controlled. The unauthorized use of a plant variety which carries a particular patented DNA sequence or which is the result of a patented technique is considered to be an infringement. Breeders cannot use the plant variety for breeding purposes, while farmers are not permitted to save seed from the actual plant to resow their land, let alone to exchange it.

It is essential that the effect of the patent is not limited to one specific variety. The patent offers the holder the right to prevent the use of all plant varieties that carry the particular patented DNA sequence or are the result of the patented technique. Several of the U.S. patents on plants provide protection covering all transgenic plants of an entire species[16], or even protect all plants in whatever species that are produced through the protected technology. For example, the USPTO has issued patents which cover all transgenic plants with increased lysine content; all transgenic plants that express higher levels of chitinase (a natural enzyme in plants that wards off fungal diseases); all sunflower plants and seeds exceeding 80% oleic fatty acid content and identified secondary trait, no matter how achieved; all edible fruit, seed and vegetables of transgenic plants that express super sweet thaumatin or monellin genes derived from African plants; all transgenic crops that are developed to express the *Bacillus thuringiensis* (Bt) toxin (a natural insect resistant); all sunflower products with low levels of saturated fatty acids; all transgenic plants that contain Bt genes using the Agrobacterium transformation method; and all transgenic plants from the *cucubitaceae* and *solonaceae* families containing cucumber mosaic virus coat protein gene (WL strain) (RAFI, 1995a:9). Prominent was the patent granted in 1992 to the biotech firm Agracetus on all forms of genetically engineered cotton. Because of its controversial character, this 'broad patent' on cotton was revoked late in 1994, although it remains valid until all the company's appeals are exhausted (Hamilton, 1996:90).

The lowering of the legal barrier to patenting plants in Europe would take a little longer. As mentioned in the previous chapter, the European industry and governments in Europe used to oppose patent coverage for plant varieties. The 1973 European Patent Convention (EPC) therefore explicitly excluded 'plant varieties' and 'essentially biological processes for the production of plants' from patent protection. Now that the wind blew from another direction, and the biotechnology industry and governments considered patent coverage for plant-related inventions to be necessary, the exclusionary provisions in the EPC constituted a problem. Two strategies were followed to make genetic information, and the plant varieties in which this information is incorporated, patentable in Europe. The first strategy involved the elicitation of jurisprudence from the European Patent Office (EPO); the second aimed at a legislative change in the national laws of the member states of the European Community.

In an attempt to circumvent the exclusionary provisions of the EPC, the attorneys defending the interests of the genetic engineering corporations aimed to narrow down the meaning of the terms 'plant variety' and 'essentially biological process'. In this way, the pool of non-patentable plant-related inventions could be reduced. The industry had its first success in 1983, when the Technical Board of Appeal (TBA) of the EPO in its decision T49/83 held patentable a chemically treated plant, produced by Ciba-Geigy. The TBA made a distinction between the term 'plant varieties' and the broader concept of 'plants'. A plant variety would refer to a group of plants that are distinct from other groups of plants, that are sufficiently uniform, and whose main characteristics remain stable after repeated propagation. Plant varieties formed the subject matter of the PVP system and were thus excluded under the EPC. Other categories of plants, however, were not. In Decision

T320/87, five years later, the TBA went a step further. At issue was a patent involving a method of producing hybrid plants, applied for by the American firm Agrigenetics, owned by Lubrizol Genetics. The TBA decided that not only the method, but also the hybrid plants it produced were patentable. Hybrids lack stability and were therefore not considered to be a plant variety. Furthermore, the Board argued that the Lubrizol method was not 'essentially biological', because the sequence of test-crossing, selection and cloning neither occurred in nature nor corresponded to the classical breeders' processes. The patentability of plants in Europe was subsequently further facilitated by Decision T19/90 of the TBA in 1991 involving the so-called 'onco-mouse', created by the Harvard University. The TBA refused to accept an earlier EPO decision in which the patentability of the mouse was rejected on the grounds that animal varieties were not patentable under the EPC. The TBA concluded that the EPC only excluded certain categories of animals from patent protection, not animals as such. Ever since, a similar reasoning was followed as regards plant varieties, which opened the door to the patenting of transgenic plant varieties in Europe, at least for several years.

To the complete surprise of many, the EPO in 1995 changed its course as regards granting patents on plants. In its Decision T0356/93 the TBA unexpectedly rejected the patentability of transgene plants in the case of the Belgian firm Plant Genetic Systems (PGS). The patent involved an invention to make plant cells resistant to glutamine synthetase inhibitors. On the grounds that all examples provided by PGS for applying its invention referred to plant varieties, the TBA concluded that PGS claimed protection for plant varieties, which are excluded under the EPC. Ever since, the patenting of transgenic plants and seeds with the EPO has no longer been possible. Early in 1999 the EPO had not yet reconsidered its position.

The PGS decision of the EPO posed a problem to the genetic engineers and European legislators, especially in view of the second strategy that had been followed to lower the legal barrier for the patenting of plant-related and other biotechnological inventions in Europe. In order to support the biotechnology industry, the Commission of the European Community (EC) had submitted the so-called 'biotechnology directive' in 1988. The EC has no direct relationship with the EPC, but all EC members adhere to the EPC. By obliging EC members to adjust their national legislation to bring it in line with the directive, the Commission could induce a legislative change in patent law in Europe without revising the EPC itself.

The Commission's proposal met with fierce political opposition from the European Parliament and many European civil pressure groups, and was adopted only in 1998, ten years after the release of the first draft. In respect of the protection of plant-related inventions, the 'biotechnology directive' basically follows the jurisprudence developed at the EPO until 1995. Plant varieties, subject matter of PVP, are excluded from patent protection, but plants can be covered by a patent "if the technical feasibility of the inventions is not confined to a particular plant ... variety." (Commission of the EC, 1998).

The biotechnology directive has two major implications on the opportunities for plant patenting in Europe. The first is that it might encourage the EPO to grant patents on transgenic plants again. This may be achieved by means of new jurisprudence or by 'implementing regulations', i.e. a new interpretation of the exclusionary provisions of the EPC in order to bring them in line with the provisions of the EC directive, without changing the provisions as such. The second implication of the biotechnology directive involves the scope of protection. In order to ease farm opposition to the directive, small farmers in Europe retain the right to propagate patented plant varieties for resowing their land. Thus, patents covering plants and seeds will not provide complete protection against on-farm propagation.

In sum, both in the USA and Europe, the opportunities to patent plant-related inventions have been greatly increased since the 1980s. Inventive techniques to produce novel genotypes and DNA sequences can be patented. In the USA, patent claims covering whole plants and plant varieties are accepted as well, while in Europe presently not.

5.3.3 PVP strategy of European plant breeding firms

The new opportunities to obtain patent coverage for plant varieties have provoked considerable opposition within the crop development sector itself, although mainly in Europe. International lobbying organizations for the plant breeding industry, such as ASSINSEL and the association of European plant breeders, COMASSO, have opposed patents with a scope of protection that exceeds fragments of genetic information on the grounds that such a broad coverage would eliminate the breeders' exemption (Van Wijk, 1998). The plant breeders maintain that those parts of the varietal genome which are not part of the patented characteristic must be freely available if the breeder succeeds in crossing out the protected fragment from the plant. Whereas ASSINSEL advocated patent coverage for plant varieties in the 1950s, in the 1990s the organization defends the breeders' exemption, which is considered to be indispensable for plant breeding. Breeding is always based on varieties that already exist, and therefore requires a broad range of variability and the free use of material (Lange, 1997:29-30). The plant breeding organizations therefore followed an alternative strategy to restrict the triple-use value of plant varieties, which would be more compatible with the requirements of plant breeding. To achieve this, they initiated a revision of the UPOV Convention, which resulted in the adoption of a new UPOV act in 1991. The new act came into force in 1998.[17]

One of the two amendments to PVP in the 1991 UPOV act that is relevant to the present discussion involves a new definition of the term 'plant variety'. With the intention to regain lost ground to the genetic engineering companies, the breeding associations challenged the narrow definition of the term 'plant variety' which was put forward by the European patent courts in the 1980s. In the new UPOV act a plant variety is therefore defined much more broadly than before[18] and refers to a plant grouping that can be defined "by the expression of the characteristics resulting from a given genotype", or as a plant grouping that can be considered as "a unit that can be propagated unchanged" (UPOV, 1991). With the subject matter of the PVP system broadened, it may be more difficult to claim patent coverage for whole plants that incorporate patented genetic information.

The second significant change in PVP concerns the freedom to use a protected variety to create a new one. Rather than eliminating the breeders' exemption, the new UPOV act restricts it by means of the introduction of the "dependency principle" for "essentially derived varieties". Dependency refers to the situation in which a new variety 'B' can only be commercialized with authorization of the title holder of the original variety 'A' from which variety 'B' is derived. However, under the 1991 UPOV act, dependency only applies if the derived variety is almost identical to the original variety. Variety 'B' is then considered as essentially derived from 'A'. Thus, the breeders' exemption remains the core of PVP, albeit that authorization of the original breeder is required when a new variety is essentially derived from an existing, protected variety. However, the question of how to determine whether or not a variety is essentially derived is yet to be resolved.

It is not entirely clear how the different IPR strategies of the genetic engineers and the plant breeders should be assessed. It seems that the emergence of crop development conglomerates has induced a split in the industry. The plant breeding firms have evolved from the farm sector and often have a long, national tradition. Now that many of them have been integrated into the crop development conglomerates, they are serving the strategies of the core industrial corporations, which seem to have adopted genetic engineering as the basic tool for crop development. The insistence on the protection of plant varieties under PVP by the breeders' associations may be an expression of some type of defence desired by breeding firms incorporated in the conglomerates. It may also be the strategy of breeding firms that still operate independently and are trying to secure free access to protected plant varieties for breeding purposes in the future. If the latter is true, and if the tendency for breeding firms to be overtaken by large industrial corporations continues, the UPOV-PVP system may lose its significance during the Third Agro-Food Order, at least in OECD countries.

5.3.4 Legal plant protection 'exported' to developing countries

While IPR protection for plant material has been a political issue in industrialized countries since the beginning of the 20th century, it is new for developing countries. Throughout the Second Agro-Food Order, in none of the developing countries plant were varieties protectable under IPR legislation. This was in part due to the fact that IPR protection was generally controversial in the Third World. There was a particular mistrust of the patent system, on the grounds that it predominantly favoured transnational enterprises, which used the patent monopoly to restrict the diffusion of their technology. This was the reason for developing countries to call for a revision of the patent system in the 1960s and 1970s (United Nations, 1975).

Until the 1990s, the absence of legal protection for plant varieties in developing countries was not considered to be a real impediment to trade by transnational breeding firms (Wierema, 1989). The industry exported hybrid seeds or relied on contracts and gentlemen's agreements when it licensed out the multiplication of plant varieties. In anticipation of possible losses due to the lack of legal protection, transnational breeding firms often demanded higher royalty rates than in OECD countries. They usually also avoided sending high quality propagating material and their latest plant varieties to developing countries (Jaffé and Van Wijk, 1995:63).

In the Third Agro-Food Order, however, developing countries are changing their position on legal plant protection drastically. Apart from domestic demands for plant-related IPR protection - which we discuss later - the changes in IPR protection have particularly been induced by foreign pressure on developing countries. The crop development conglomerates, core actors in the design of the world's new commercial plant varieties, not only provide the capital, technology and genetic information, but also bring along their regulatory demands. Basically, they have used two strategies to 'export' legal instruments for plant protection.

The first strategy has been increasingly to make plant variety or patent protection a condition for access to their high quality propagative material of advanced plant varieties. Producers in developing countries that are involved in export crops, such as maize, soyabean, fruit or ornamental plants, therefore have an interest in accepting these legal requirements. Except for some of the largest developing countries, the plant varieties that are required for export are seldom bred domestically. They are normally obtained from the foreign crop development industry and subsequently adapted and propagated locally. The cultivators of these export crops have therefore been the main domestic advocates of legal plant protection in a number of Latin American countries (Jaffé and Van Wijk, 1995:30-32).

The second strategy to have legal plant protection introduced in developing countries has been the political lobby of the crop development conglomerates during international trade negotiations. In the 1980s, it was realized that economic growth greatly relied on the innovative potential of knowledge-intensive industries, based on computer technology, new materials and biotechnology (*cf.* Dosi *et al.*, 1988; Roobeek, 1990). The protection of this knowledge as intellectual property was considered a strategic element in the defence both of the domestic industry and of the competitiveness of the national economy. In this way, PVP and patent protection for genetic information were not treated as isolated issues, but as part of an overall strategy towards a stronger, worldwide IPR protection in all areas of technology.

In the early 1980s, a powerful coalition of U.S. private industry associations in advanced technological sectors managed to make the international protection of intellectual property a priority issue in the USA. The coalition, representing the pharmaceutical, chemical, computer and information processing, aerospace, and entertainment sectors, lobbied successfully for a 'trade-based' intellectual property approach. It laid the blame for the growing U.S. trade deficit on foreign (developing) countries' unfair practices and offered an alternative to mounting protectionist pressure at home. The private exporting industries that were represented in intellectual property lobbies came from industrial sectors that enjoyed positive trade balances. They could present their industries as a solution to America's trade problems (Sell, 1995:165,178). Intellectual property began to be considered "America's overlooked export" (McCarthy, 1995), and "our once and future strength" (Oman, 1994).

In response to the private intellectual property lobby campaign the White House, in the early 1980s, announced a new policy on intellectual property[19], while U.S. Congress passed into law a series of amendments to the U.S. Trade Act. In doing so, Congress made adequate IPR protection a condition for foreign countries to have trading and investment relations with the USA. The amendments, especially those concerning Section 301, enabled the U.S. government to take retaliatory measures against trading partners that did not comply with American intellectual property demands (Van Wijk and Junne, 1993).

Apart from prompting the U.S. government to force other countries to protect American intellectual property better, private industry paved the way for the negotiation of a multilateral intellectual property agreement in the Uruguay Round of GATT (Pratt, 1995). GATT made package deals possible in which trade benefits in other areas were linked to IPR protection performance. Moreover, the dispute settlement mechanism of the GATT could be used to facilitate action that could be relatively quickly enforced against countries violating intellectual property regulations.

Thirteen major U.S. corporations formed the Intellectual Property Committee (IPC)[20], dedicated to a comprehensive IPR agreement in GATT. The IPC allied itself to the business communities in Europe and Japan and convinced them that an international agreement was possible. As a 'Trilateral Group', they outlined the fundamental principles for protecting all forms of intellectual property (IPC *et al.*, 1988).

In 1994, most of the 125 participating governments signed the agreement on Trade-Related Intellectual Property Rights (TRIPS), which was part of the new GATT. The industry hailed the TRIPS agreement as a great success.[21] In crop development, the trilateral industry group largely obtained what it wanted. The agreement obliges member states, *inter alia,* to adjust their patent laws and to grant patent protection for any invention in all fields of technology. Inventions related to biological materials, such as genes and DNA sequences, are not explicitly excluded from patent coverage and are therefore considered to be patentable. Plants and plant varieties may be excluded from protection. Plant varieties, however, are at least protected by an effective *sui generis* system, which implicitly refers to the UPOV-PVP system. The fact that legal protection of plant material was controversial during the negotiations is illustrated by an additional provision which says that the opportunity to exclude plants and plant varieties from patent protection is to be reviewed four years after the agreement has come into force.

In sum, the quest for broader and more effective IPR protection for plant genetic material and plant varieties worldwide has resulted in most countries accepting patent protection for genetic information. Some of them, such as Colombia, also allow patent protection for plant varieties as such, and essentially biological processes for the production of plants. As of 1998, eight developing countries[22] have adopted PVP legislation and acceded to UPOV. UPOV's total membership increased from 21 countries in 1992, to 37 in March 1998 (UPOV, 1998:33).

5.3.5 IPR in support of a new international division of labour in crop development

What we have seen above is how the changing international division of labour in crop development initiated a worldwide adjustment of IPR protection for plant-related innovations. The question pending is how the new IPR rules will support the position of the crop development conglomerates during the Third Agro-Food Order. Will the new worldwide minimum standard for plant-related IPR protection indeed sustain the oligopolistic tendencies in crop development? Will it also offer opportunities for the domestic crop development industry in developing countries? Will IPR protection deprive farmers from opportunities to produce their own seed, or will they benefit from the advanced technology incorporated into the new plant varieties offered by the conglomerates?

It is too early to provide conclusive answers to these important questions. Patent protection for genetic information is a new phenomenon and primarily confined to the OECD countries. Moreover, knowledge of the socio-economic effects of IPR protection on developing countries is almost non-existent (Siebeck, 1990). Despite these hindrances, we will provide some indicative answers. The effects of patents cannot yet be measured, but may be estimated on the basis of experience with PVP, which has been functional for around 20 years. Here we have to cope with another limitation, however: the literature on the effects of PVP is extremely scarce. Only a very few publications exist. We derive our information on the impact of PVP in developing countries basically from a study conducted in the mid-1990s in five Latin American countries (Jaffé and Van Wijk, 1995; Van Wijk and Jaffé, 1996).

(a) Exploiting genetic information for breeding

It is obvious that the crop development industry in OECD countries is benefiting from the new worldwide IPR regulation, which the industry itself demanded. The more countries adopt PVP legislation, the larger is the geographical area within which new plant varieties are protected from unauthorized multiplication. For the genetic engineering corporations, which aim at patent coverage for plant varieties, PVP also is beneficial as it clears the ground for more public acceptance in developing countries of patenting plants in the future.

The data on PVP applications and grants in some Latin American countries show that the transnational breeding firms have immediately used the new PVP opportunities. In Argentina and Chile, until 1994, around one-third of PVP titles were granted to foreigners, chiefly on alfalfa and soyabean varieties and on inbred lines that are the parents of maize hybrid varieties (Argentina), as well as on fruit and potato varieties (Chile) (Jaffé and Van Wijk, 1995:39). The first wave of PVP applications in Colombia in 1997 almost exclusively involved foreign organizations, which applied for protection mainly of cut flower varieties (UPOV, 1998:9). PVP obliges producers in Argentina, Chile and Colombia to request authorization for the commercial propagation of protected varieties, to pay a royalty, and to agree to the (export) restrictions usually imposed in the licensing contract by the foreign breeder.

Apart from supporting plant breeding firms in OECD countries, PVP legislation may also have a positive influence on domestic plant breeders in non-OECD countries. In Argentina, PVP has presumably played a supportive role for domestic private investment in wheat and soyabean breeding, or has at least prevented a reduction of such investment (Jaffé and Van Wijk, 1995). This effect is similar to that which has been found in the USA (Butler and Marion, 1985; Butler, 1996). PVP also seems to support public breeding institutes in a number of Latin American countries. Budgetary pressures are forcing these institutes to take advantage of PVP as a means of developing an alternative source of revenue.

The protection of genetic information and plant varieties by patents is, as yet, confined to OECD countries. We can expect, however, that patents generally sustain the position of the OECD-based crop development conglomerates vis-à-vis independent plant breeding firms in both OECD and non-OECD countries. A patent covering a plant will usually be obtained by those organizations that have mastered genetic engineering techniques, and that can afford the patent defence. The latter especially cannot be underestimated. It is common knowledge within patent law circles that the strength of a patent depends largely on the financial strength of the patent holder.

Patents on genetic information and plant varieties are likely to be held by the life science corporations that guide the crop development conglomerates, rather than by small, independent plant breeding firms. In none of the countries where the patent has been granted are breeders allowed to use the protected genetic information or plant variety. Since a breeder's exemption is absent in patent law,[23] the use of patented genetic information or varieties must be authorized by the patent holder through a licence contract. Such a contract usually requires a royalty payment, and stipulates to what end the genetic information can be used, in what quantity, and in what countries the new plant variety may or may not be marketed.

Since corporations integrated in crop development conglomerates tend to protect technology and genetic information as much as possible, the breeder of a new variety, who must build on existing techniques and varieties, finds himself confronted with a complex 'IPR pedigree'. Seghal (1996) describes such a possible pedigree in relation to the development of a new transgenic, insect-tolerant plant variety, which "may involve plant variety rights, patents on plants, as well as several patents relating to transformation technology, the selectable marker employed, the gene coding for the insecticidal protein, the promoter, and various regulatory elements and modifications needed to express genes adequately in plant cells.". When the use of only one element of this required technology package is denied by an IPR holder, the commercialization of the entire variety may be blocked.

The prime option for crop development corporations to cope with this interdependence on protected technology is to swap their intellectual property rights. The disadvantage of swapping is, however, that it reinforces the oligopolistic tendencies in the crop development sector. IPR swaps raise barriers for entry into crop development, as only firms that possess an IPR portfolio may have access to proprietary breeding technology in the future. The small independent breeding firm, most common in non-OECD countries, is likely to encounter difficulties in obtaining licences for using protected innovations. Some local seed firms will obtain a licence, because they can provide the locally adapted plant varieties that will serve as carrier for foreign genetic innovations. According to Dekalb CEO Bickner, local private seed companies will become "regional product testers and merchandisers", giving the local service to farmers (Freiberg, 1997a:6). Patent law, however, allows such a licence to be exclusive, preventing other breeders in the country to use the same plant and genetic innovation. They may request their national government to force the patent holder to license his technology on a non-exclusive basis, by means of a so-called 'compulsory licence'. This instrument, however, is only applicable in extreme cases of public interest and in practice seldom used (Van Wijk and Junne, 1993:33-35).

In sum, while taking into account all disclaimers regarding the meagre evidence that exists, the worldwide strengthening of legal protection of plant-related innovations seems basically to support the OECD-based crop development conglomerates. Although PVP may have a positive effect on domestic private plant breeding firms and semi-public institutes in non-OECD countries, it is difficult to see how crop development in these countries is facilitated by the granting of patents on genetic information and plant varieties.

(b) Exploiting genetic information for propagation

An assessment of the impact of legal plant protection on the international division in crop development cannot be confined to the crop development sector alone, but must also cover the consumers of seed: the farmers. A customary practice in agriculture is that farmers produce their own seed, swap saved seed with other farmers, or swap saved seed for grain. This practice tends to decline with the industrialization of agricultural production. Large and capital-intensive farms in OECD countries use little on-farm saved seed, but the vast majority of seed used in developing countries is on-farm saved, and/or supplied through informal distribution channels (Tripp, 1997:18). The replacement of this informal seed exchange by a formal market of commercial varieties is an attractive prospect for the crop development industry, as it increases seed sales. The industrial marketing initiatives fit the prevailing political strategies in developing countries to industrialize agriculture further.

Legal protection of genetic information and plant varieties substantially promotes the replacement of traditional varieties by advanced ones. The implementation of PVP in Argentina significantly reduced informal seed exchange of protected varieties of wheat and soyabean between 1990 and 1994, and recent experiences in Chile indicate similar results (Van Wijk, 1997). In Argentina, the costs of PVP enforcement have initially been paid by those seed dealers that were accustomed to selling and swapping seed for grain without authorization and without paying royalties and taxes (Jaffé and Van Wijk, 1995). In due course, however, the higher costs of the seed dealer will probably be passed on to the farmers. Seed prices may rise considerably, as seed is increasingly priced for the specific genetic information it contains. In the USA, 'technology premiums' of US$ 75 per hectare of cotton and US$ 25 per hectare of maize of Bt-based insect protection have been reported (Seghal, 1996).

The restrictions on seed saving are tighter under patent law. While on-farm saving of seed for private resowing purposes is usually allowed under PVP, it is not under patent law. Only in Europe are "small farmers" placed in an exceptional position. According to the EC biotechnology directive of 1998, they are allowed to save seed of plant varieties protected by a patent. To European farmers who are not considered to be "small" as well as to other farmers in the rest of the world, patent coverage for plants has more detrimental effects in terms of seed saving than PVP.

It is likely, however, that in the future, new technologies specifically developed to restrict unauthorized propagation will affect farmers more than patents. Many farmers already grow hybrid varieties, which have built-in protection against propagation, while recently new mechanisms to prevent unauthorized propagation have been developed. Widely publicized has been the 'technology protection system' (TPS), jointly developed by the USDA and Delta & Pine Land, the world's largest cotton company and part of the Monsanto conglomerate. TPS, referred to by critics as the "terminator technology", modifies the DNA of plant varieties with the aim to make on-farm saved seed sterile in order to prevent it from germinating. TPS is being tested in cotton and tobacco, and is claimed to have broad applications in other self-pollinating crops in which commercially viable hybrid varieties are not yet developed, such as wheat, soyabean, rice, and oats. According to USDA spokesman W. Phelps, the technology has been expressly created to "increase the value of proprietary seed owned by U.S. seed companies and to open up new markets in Second and Third World countries" (RAFI, 1998b:2). In Europe, the British company Zeneca BioSciences is also working on a technology that discourages unauthorized on-farm propagation of plant varieties. The technology activates a "killer" gene, or prevents the expression of genes crucial to normal plant development, and weighs in whenever a chemical "trigger" is applied to seed at a desired point in plant maturation. The bought seed will either not rejuvenate or will not function in subsequent generations (RAFI, 1998c).

Considering that in most developing countries a large part of the rural population depend on-farm saved seed, it is hardly surprising that it is especially in these countries that the worldwide tendency to restrict the dual use of plants is opposed. It is equally unacceptable to plant breeders in developing countries that foreign transnational enterprises are enabled to prohibit others the use of their plant varieties for creating new ones. It is this concern that intensifies the international conflict on legal plant protection, a conflict that is dealt with extensively in the next chapter.

Conclusions

The advent of the Third Agro-Food Order has revealed a tendency for the state as the pivot of crop development to be replaced by private industry. Since the 1980s, the growth of public investment in agricultural R&D has declined, private industry has obtained a greater say in the allocation of public agricultural R&D funds, while private investment in agricultural research has risen rapidly. This development has been accompanied by a thorough restructuring of the organization of the plant breeding sector, which has given rise to the formation of industrial crop development conglomerates, based in OECD countries. Given the accumulation of unrivalled financial and technological capacity within these industrial conglomerates, they seem to become the central actors and dynamic force of crop development in the Third Agro-Food Order. The crop development conglomerates have induced two major changes in conservation strategy and IPR regulation worldwide:

(a) A broadening of the conservation target from landrace seed to biodiversity
The vast investment by the industrial conglomerates in crop development has been greatly facilitated by the technical capacity to replace the plant variety by DNA base pairs as the main object of research. This replacement of plants by genetic information has considerable consequences for conservation strategies. Whilst previously seed of landraces was collected and conserved on the basis of their phenotypical characteristics; in the Third Agro-Food Order the collection is increasingly guided by genetic criteria. Genetic information of any organism - be it a microorganism, a plant, an insect, or an animal - may be the target of collection and conservation, as long as the information codes for a commercially valuable trait to be incorporated in a plant variety. The broadening of the genepool for the industrial conglomerates opened the door for joint biodiversity conservation initiatives by the industrial and non-governmental environmental protection organizations.

(b) The quest for elimination of the unauthorized exploitation of genetic information for breeding and propagation
The opportunities of genetic engineering have rendered inadequate the existing systems for the protection of intellectual property of breeders. Both PVP and hybridization aim to reduce the unauthorized use of a plant variety for propagation; they do not, however, prevent the use of the genetic information incorporated in these varieties for the generation of new varieties. For this reason, the genetic engineering companies of the conglomerates seek to protect plant genetic information under the patent system. Patents protect the genetic engineer from all unauthorized exploitation of the genetic information and all varieties - including the progeny - which contain that information. The prohibition of the use of patented plant varieties for breeding has aroused opposition, especially from European plant breeding associations. This reaction, however, may be interpreted as rear-guard action of a vanishing type of company, the independent plant breeding firm.

The crop development conglomerates have their homebase in OECD countries but intend to supply the global market. Increasing demand for commercial seed in developing countries, in combination with decreasing public investment in agricultural R&D, induces commercial agriculture in developing countries to rely on varieties produced by the crop development conglomerates. This development indicates that a new division of labour emerges in crop development itself, with the industrial conglomerates as the world's main suppliers of plant varieties.

In this chapter it has been argued that some major adjustments in the prevailing conservation strategy and IPR regulation have been made to facilitate this new division of labour. In order to be able to supply the entire world with improved plant varieties, the crop development conglomerates have advanced the strengthening of plant-related IPR protection through the agreement on TRIPS. The Convention on Biological Diversity, in turn, sets out new rules for the access to the diverse biological organisms, containing genetic information.

At present, we are only able to identify the early stages of tendencies that seem to shape the Third Agro-Food Order, and it is only with caution that we have made extrapolations well into the next century. We suggest that the division of labour in crop development will strongly support the internationally dominant position of OECD countries in this branch of industry. The vast private and public expenditure on mapping the genome both of crop plants and of the worldwide collection of diverse biological organisms is primarily directed at the creation of DNA databanks, which will be of strategic importance to the competitiveness for crop development and the agro-food industry of OECD countries. Seedbank collections consequently have lower priority for crop development in OECD countries, but remain important for plant breeding in developing countries.

Access to DNA data, and to the new generation of plant varieties generated by genetic engineering, is increasingly restricted. Even if advanced DNA data could be applied by the plant breeding industry in developing countries, patent protection would effectively prevent any unauthorized use for breeding. The present introduction of PVP, however, is of limited use for the protection of transgenic plant varieties, and thus for the crop development conglomerates, but may be more useful for independent private breeding firms, including those in developing countries.

Notes

[1] The assumption that specific DNA-sequences or genes are responsible for specific plant characteristics is not uncontroversial. It has been criticized as a reductionistic understanding of the functioning of living organisms. Kollek (1995:24), for example, states that it is not just a gene's sequence that determines its functions in the organism, but also its location in a chromosomal, cellular, physiological and evolutionary context.

[2] The OECD shares are higher when the value of import and export of agricultural products is taken into account. Figures 5.1 - 5.10 present agricultural trade in volume because this offers a better picture of real changes in trade flows.

[3] The main exporters of NTCs are China (US$ 1.42 billion in 1994), Mexico (US$ 1.26 billion in 1996-

97), and Thailand (US$ 738 million in 1996-97) (FAOSTAT, 1998).

4 The exact value of the commercial seed market is difficult to measure. The interpretation of the terms 'seed and plantlet' varies from one country to another, detailed statistics are absent, and the amount of farmers' saved seed differs considerably per country and is difficult to evaluate (Le Buanec, 1996:409). For this reason, the estimated commercial world seed market ranges from US$ 15 billion (Heijbroek *et al.*, 1996) to US$ 28 billion (Le Buanec, 1996). Because the value of the seed market is rising rapidly due to the higher prices for transgenic and specialty seed, the highest estimate is likely to be the most accurate.

5 Calculations of agricultural R&D must be used with some caution. According to Alston *et al.* (1998:53) there is no single, internationally agreed notion of what constitutes 'agricultural R&D'. The authors themselves use a broad definition of agricultural research which includes research on "crops, livestock, forestry, and fisheries and related environmental R&D", targeted at "pre-production (e.g., input supply), on-farm production, and post-production (e.g., food processing) technologies". It is unclear, however whether, or to what extent, public funds for genetic engineering R&D have been included. In view of national competitiveness, governmental expenditure in biotechnology became a recognized tool of OECD countries in the 1980s to support advances in molecular biology. Even though most of this investment was earmarked, either as 'health', 'food' or 'agriculture', the partial integration of the R&D for these sectors into 'life science' R&D, renders the term 'agricultural R&D' increasingly inapt.

6 The data on private agricultural R&D provided by Alston *et al.* (1998) are derived from OECD's intramural business sector series and are calculated as the sum of R&D performed by the industry classified as "agriculture, forestry, fisheries", and "food and beverages". Added to these figures is 10% of total research performed by the chemical and pharmaceutical industries. This proportion of chemical R&D is considered to be assigned to agriculture on the basis of an "approximate rule of thumb procedure that was chosen after consulting various other relevant sources" (Alston *et al.*, 1998:66).

7 There are in principle two approaches to calculate private R&D in crop development: as a proportion of total private agricultural R&D, or as a proportion of total turnover of the private crop development industry. The first approach is hindered by the variation in defining the term agricultural R&D and sub-categories of this R&D. For the USA, for example, it has been calculated that in 1992, 12% of total private agricultural R&D was spent on "plant breeding" (USDA, 1995). The accuracy of this percentage must be doubted, however. The corresponding amount of money spent in this category, US$ 400 million, seems incompatible with the R&D funds allocated by the crop development conglomerates in the mid-1990s. Figure 5.14 shows, for example, that the 1998 R&D expenditure of Monsanto alone amounted to over US$ 1 billion. We prefer, therefore, the alternative approach, in which private R&D in crop development is calculated as proportion of turnover. It has been indicated that R&D expenditure in crop development is between 5 and 20% of total turnover (Heijbroek *et al.*, 1996:48), or 15% (Thirtle *et al.*,1997:52). If we assume the R&D proportion to be between 10 and 15% of the total commercial seed market, which we have earlier estimated at around US$ 28 billion, total private crop development R&D would be between US$ 2.8 and 4.2 billion. These figures seem more in line with the private research expenditures shown in Figures 5.14 to 5.19.

8 There were also other reasons for breeding firms to integrate with the large chemical corporations. In The Netherlands, several breeding firms run the risk of rupturing because third generation family members (the grandchildren of the founders) preferred to withdraw their shares from the family firm. A merger with a chemical corporation was considered to be a solution to this situation. (Personal communication N. Louwaars, September 1998, Centre for Plant Breeding and Reproduction Research, Wageningen, The Netherlands)

9 The recent involvement of the Mexican corporation Empresas La Moderna (ELM) in crop improvement

is a good illustration of this tendency. In a short period, ELM acquired a number of European and American vegetable-breeding firms. In this way, the company has become the world's primal player in genetically engineered vegetable seeds, vegetable production, and distribution. The company has set up a system of vegetable contract farming on the basis of ELM seed (Massieu, 1998).

[10] Two waves of take-overs during the 1980s and the 1990s have been widely publicized. A number of chemical companies which became involved in the seed industry in the 1980s, such as Shell, Pfizer, Rohm and Haas, Elf Acquitaine, divested themselves of their seed and biotech interests a decade later because the seed sector did not fit into their corporate strategies for a number of reasons. Firstly, it took much longer to apply genetic engineering techniques in the development of new products than had originally been envisaged. Secondly, seed, unlike chemicals, could not be marketed globally but had to be adapted to local climatic conditions. Thirdly, there were conflicts in management style between the smaller (formerly family owned) seed companies and the large chemical companies (Seghal, 1996:18).

[11] See, for example, the brochures *Feeding the 5 billion* (ASSINSEL, 1981) and *Feeding the 8 billion and preserving the planet* (ASSINSEL, 1997), and the Internet homepage of Monsanto titled "Food, Health, Hope" (http://www.monsanto.com) starting with the following message: "There is family that lives here ... a family of six billion, each with the possibility of living longer and healthier through the discovery we, the people of Monsanto have just begun ...". DuPont CEO Krol comes with a similar message: "We know that the projections for population growth show a rise from 6 billion people to 10-12 billion people with the next 25-40 years ... For DuPont, a multinational corporation ... this represents an enormous opportunity to create value for society, the environment and the shareholder ...." (Krol, 1997).

[12] Opportunities to expand production through an extension of arable land are very limited, while productivity growth is presently levelling off for the main crops. In the period 1985-1995 the productivity growth of wheat, maize, rice and soyabean was around a third of the production growth during the preceding decade (Heijbroek *et al.*, 1996:11).

[13] According to the Convention on Biological Diversity (CBD) 'biodiversity' encompasses no less than "[T]he variability among living organisms from all sources including, *inter alia,* terrestrial, marine and other aquatic ecosystems and the ecological complexes of which they are part; this includes diversity within species, between species and of ecosystems." (UNEP, 1992:27). 'Agro-biodiversity' refers to the biodiversity in crop plants and animals and constitutes one element of total biodiversity.

[14] When, during the Second Agro-Food Order, private breeding gathered steam, the difficult task of extracting new, useful genetic information from uncrossed landraces was increasingly considered a task of public breeders, which worked in a less competitive environment. However, when in the 1980s public breeders were forced by decreased public funding to speed up the pace of their new releases, they too started to avoid landraces and seedbanks. In the 1980s, both private and public breeders in the USA reportedly asked for landraces only in about 5-10% of all occasions they approached seed banks (Cox *et al.*, 1988:117; Duvick, 1984b; Rejesus *et al.*, 1996:132).

[15] Patents on plant-related innovations should not be confused with the American 'plant patents' granted under the Plant Patent Act of 1930. In this publication the term 'patent' refers to 'patent for invention' or 'utility patent'.

[16] For example, U.S. patents have been issued on all transgenic plants created by specific transformation techniques of the *Brassica* family (rapeseed, broccoli, cauliflower, cabbage and Brussels sprouts), *Coffea arabica*, the most important commercial coffee species, *Cucumis melo* (melons, muskmelons and cantaloupes), *Capsicum* (pepper), and *Pisum sativum L.* (garden peas) (RAFI, 1995:9).

[17] The 1961 UPOV act was amended twice on earlier occasions. In 1972 an additional act was adopted to improve the organization of UPOV. In 1978 the UPOV act was revised, basically to allow the accession to the Union by the USA, which also granted plant patents for some type of plants. The re-

vised text adopted in 1978 differed little from the 1961 text (*cf.* Heitz, 1987:89-93).

[18] In the 1961 act of UPOV the word 'variety' generally applied to "any cultivar, clone, line, stock, or hybrid which is capable of cultivation" and sufficiently homogeneous and stable. Remarkably, in the revised text of the 1978 UPOV act a definition of the very object of protection was left out.

[19] According to former Pfizer CEO Pratt: "This was a signal to U.S. international policymakers in various departments that the President cared about IP. All of a sudden, American ambassadors in country after country become advocates - often strong advocates - for IP vis-à-vis pirate governments." (Pratt, 1995).

[20] The IPC included the pharmaceutical companies Bristol-Myers, Johnson & Johnson, Merck, and Pfizer; the computer and information-processing industry with General Motors, Hewlett-Packard, and IBM; the aerospace and defence corporations FMC Corporation, General Electric, and Rockwell International; the chemical corporations DuPont and Monsanto, and Warner Communications from the entertainment sector.

[21] Enyart of Monsanto stated: "Besides selling our [intellectual property] concepts at home we went to Geneva where we presented [our] document to the staff of the GATT secretariat. We also took the opportunity to present it to the Geneva-based representatives of a large number of countries.[...] What I have described to you is absolutely unprecedented in GATT. Industry has identified a major problem in international trade. It crafted a solution, reduced it to a concrete proposal and sold it to our own and other governments." (quoted in Sell, 1995:181).

[22] These countries include Argentina, Chile, Colombia, Ecuador, Mexico, Paraguay, Trinidad and Tobago, and Uruguay.

[23] The patent laws of most countries recognize a 'research exemption'. This exemption is far more restrictive than the breeders' exemption in the sense that it allows the use of a patented invention for non-commercial research purposes only, but there are considerable differences among countries. In the USA, authorization is usually required to use a protected invention for research purposes, while in Europe it is not.

Chapter 6

International Division of Labour in Crop Development: Implications for Chile and Colombia

In the previous chapter we examined the main tendencies in crop development that seem to give shape to the emerging Third Agro-Food Order. We have shown that a small number of industrial conglomerates is replacing the state as main organizer of crop development. They have successfully influenced the strategies for conservation and IPR protection, which have a worldwide impact. However, being the dynamic force of crop development, the industrial conglomerates meet with considerable opposition, not only in the OECD but also in developing countries.

In the international division of labour in agriculture, developing countries generally take a position that is different from most OECD countries. Overall in developing countries, the level of industrialization of the agricultural sector is lower than in OECD countries, which is reflected in a lower productivity and a larger part of the population depending on agriculture. As a result, the share of developing countries in the world's imports of some major staple crops has continuously increased in the past, while the share of OECD countries in the world's agricultural exports has risen (two tendencies that we have shown earlier).

In crop development there are also wide differences. The industrial crop development conglomerates have their home base invariably in one of the OECD countries, while nearly all genetic engineering research takes place in these countries as well. In developing countries, the main breeding organizations are generally public institutions, whose budgets have been reduced in the past decade. As the domestic private breeding industry is not very well developed in these countries, the role of national government in crop development tends to be taken over by the transnational industrial conglomerates, at least for a number of commercial staple crops. Considering these rough indications of differences in agricultural production capacity between OECD and developing countries, it is obvious that the effects of the prevailing agricultural production strategies, including specific crop development policies, will be different for both groups of countries as well.

In this chapter we examine the agricultural sectors of two Latin American countries, Chile and Colombia in more detail. We explore their specific position in the international division of labour in agriculture and in crop development. Some of the major political dilemmas in the agricultural production strategy since the 1980s are discussed. The main argument we develop in this chapter is that at the root of controversies on conservation and IPR protection lie rival views on how the agricultural sector should be organized. It is the increasing economic marginalization of the poorest rural population that constitutes the basic reason for opposition to conservation and IPR protection policies.

The chapter consists of three sections. In the first, we explain three agricultural production strategies and relate them to the various agricultural sectors and policies in Chile and Colombia. Then we focus on the question of how the various production sectors and agricultural policies are related to respectively conservation and plant-related IPR protection policies.

6.1 Conflicting interests in the organization of agriculture

Those who travel through rural Colombia will experience great contrasts in agricultural production. The sight of large, well-equipped flower production plants and extensive plantations near Bogotá is soon replaced by one of very poor subsistence farmers in more mountainous regions. Similar contrasts can be found in Chile. In this chapter we argue that the main causes for the PGR conflict must be sought in these contrasts. The present section is devoted to the question of how these large differences in agricultural production can be interpreted.

We start off with a discussion on the agricultural production strategies in the Third Agro-Food Order; strategies which refer to the more or less consistent sets of policies for the organization of agriculture. In OECD countries most agriculture follows one fairly uniform, usually highly industrialized production strategy. In Chile and Colombia, and in most other developing countries, the picture is far more fragmented: agriculture follows more than one production strategy, and the advocates of each strategy have a particular view on how conservation and IPR regulation should support agriculture.

6.1.1 Three agricultural production strategies

In the first chapter of this book we distinguish three alternative and conflicting strategies for the organization of agriculture: 'market-led' and 'state-led' agro-industrialization strategies, and a third strategy directed at non-industrial, farmer-oriented agricultural production (Figure 6.1). These rival strategies must be interpreted as ideal-types in the Weberian sense. They are an exaggeration of some essential features of agricultural production strategies of public and private organizations. The strategies encompass all socio-economic, technological and legal policies that are intended to speed up, steer, or slow down the process of agro-in-

Figure 6.1 Typology of agricultural production strategies

	Industrialization of agriculture		
	Market-led	**State-led**	**Non-industrial,farmer-oriented agriculture**
Orientation	Global	National	Local
Leading principle	Production on the basis of private industry decisions.	Production on the basis of the national political agenda.	Maintenance of agricultural production capacity in rural areas.
Role of the state	Facilitating industrialization of agriculture by supporting private enterprises.	Steering the process of agro-industrialization. Slowing down socially undesirable effects and stimulating desirable developments.	Supporting non-industrial improvements of agriculture.
Main policy objective	National specialization. Production on the basis of competitive advantage. Unhindered agricultural trade.	Strengthening and defending national agricultural production capacity.	Securing non-market oriented, small-scale agricultural production. Maintenance of natural environment and agro-biodiversity.
Justification	"Feeding the world". Securing global food security.	Securing national food security, both quantitatively and qualitatively.	Securing local food security.
Conception of nature	Nature protection through selective protective measures and more advanced technology.	Nature is to be exploited on the basis of the national political agenda.	Nature as material and cultural basis for survival.
Focus of crop development	Raising productivity and quality of those crops that have international, competitive potential, by means of advanced technology.	Raising productivity and quality of those crops that are important for national farmers and consumers by means of advanced technology.	Raising yield security of endemic crops by utilizing farmers' knowledge, and locally available natural resources.

dustrialization, or to support non-industrial agriculture. The typology is helpful to understand the political differences of opinion about how agriculture should develop and, consequently, what crop development policy should be followed.

The first strategy is directed towards a *market-led industrialization* of agriculture, i.e. the transformation of natural and farm processes into technical and industrial processes at a pace and along lines set by the market. The state has a facilitating role in speeding up the transformation process. The strategy is global in its orientation. The world is perceived as if it were one global agro-food enterprise in which production and distribution are effectively to be organized through the market. To this end, the agricultural sectors of the individual countries should specialize and compete on the basis of their comparative advantages, while agricultural trade should be unhindered by national political interference. National crop development should be based on all the scientific information that is internationally available. It should focus on the improvement of productivity and quality of those crops that can be produced nationally against international competitive prices. Research on crops that can be imported at lower costs should be avoided. In the market-led production strategy a prominent position in "feeding the world" is being attributed to the transnational agro-industry. Concern for environmental degradation is directed to the "carrying capacity of the earth".[1] Genetic engineering is expected to play a major role in making agriculture more sustainable.

The second strategy is directed at a *state-controlled industrialization* of agriculture and implies a prominent role of the national state in the organization of agricultural production. Governmental involvement is led by the national political agenda, in which a broad variety of national producer and consumer interests are taken into account. The steering role of the government may take two directions. On the one hand, some developments in agro-industrialization may be slowed down in order to avoid social costs, such as the too rapid elimination of farms, the risk of national food shortages, or concern about health aspects of genetically modified food. On the other hand, governments may encourage politically desirable industrial developments by funding agricultural research, offering tax benefits, etc. In the state-led strategy, crop development is focused on raising the productivity and quality of those crops that are considered to be important for national farmers and consumers. Internationally available and nationally accepted technology is utilized to this end. Nature is valued as a source of economic development, which should be exploited according to the national political agenda.

The third strategy is directed at a *non-industrialized, farmer-oriented agriculture*. It is the strategy that focuses on the maintenance and improvement of small-scale farming systems, and is in support of farmers who have been marginalized during the process of agro-industrialization. This strategy aims at more farmer autonomy in agricultural production by avoiding dependence on the use of external, industrial inputs. The maintenance of rural agricultural production capacity is considered to be essential for reasons of local food security. Crop development must be directed to specific local needs and tastes, as well as from native crops and other locally available natural resources. The concept of nature seems to be ambiguous within this strategy. Nature and farm-oriented NGOs in developing countries who support this strategy tend to value nature as the material basis for life and survival, and as the spiritual environment for culture and religion. Nature and farm-oriented organizations in OECD countries, however, tend to value nature as the eco-system on which humanity depends and as the antithesis of the urban, profit-led civilization. Farming communities that operate in semi- or non-industrialized production systems, especially those located within 'centres of genetic diversity', are valued as custodians of the world's biodiversity.

The conflicts between the respective strategies are relatively modest overall in OECD countries compared to developing countries. As most agriculture in OECD countries is integrated into the agro-industrial market economy, political support for non-industrialized agriculture is marginal. The support for this strategy can mostly be found with nature and farm-oriented NGOs, and donor organizations that focus on non-industrialized agriculture in developing countries. In OECD countries, the political controversy is basically between the market-led and state-led strategies. Judging from publications and Internet homepages, the market-led strategy is strongly promoted by, for example, transnational agro-industry organizations, such as the International Agri-Food Network[2] (IAN) (IAN, 1998), the Biotechnology Industry Organization (BIO) in the USA (BIO, 1998), and its European counterpart EuropaBio (EuropaBio, 1998), and also by organizations such as the World Bank (Barghouti *et al.*, 1993) and the International Chamber of Commerce (ICC, 1997).

We indicated in the previous chapter that the market-led agricultural production strategy has prevailed in OECD countries since the 1980s. It has mainly been challenged in Europe, where farm and consumer organizations have tried to steer and slow down some trends in the ongoing agro-industrialization promoted under the Common Agricultural Policy. European farmers have regularly blocked roads or dumped their produce by way of protest against the absence of more supportive and protective measures. There is also considerable resistance against the patenting of plants and animals, as was illustrated by the ten-year struggle over the EC's biotechnology directive. Furthermore, the introduction of genetically modified organisms seems controversial everywhere in Europe (Flynn *et al.*, 1998). Nevertheless, compared to the situation in some developing countries, the social conflicts that emerge in OECD countries from political choices regarding the agricultural production strategy remain relatively insignificant. Large-scale social unrest in this respect has hardly occurred. The financial compensation schemes, and alternative employment opportunities for farmers that are forced to retire prevent the social cost incurred by farmers from being too high. Furthermore, the number of people that translate their concern about developments in biotechnology into political action is relatively small.

In many developing countries, the situation is the opposite. Agriculture is usually much more important, both in terms of contribution to the GNP, and number of people who rely on the national agricultural sector for their income or food supply.

Moreover, developing countries generally hold a weaker position in the international division of labour in agriculture than do OECD countries. This mixture implies that choices related to agricultural production strategies may have a strong impact on the population, and therefore may be very sensitive politically. By way of example, we will explore the political dilemmas and conflicts over the rival production strategies in Chile and Colombia.

6.1.2 Agricultural policy in Chile and Colombia

Both in Chile and Colombia, the conflicting interests in agricultural production became prominent with the *apertura*, the opening of the national market. The *apertura*, introduced in Chile in the late 1970s and in Colombia in the late 1980s, gave way to a more market-led agricultural production strategy. In both countries, state involvement in agriculture was reduced or redirected. Import tariffs, quantitative import restrictions, price support, government procurement, and funding of public agricultural research institutes were lowered or abandoned. At the same time, national trade-related laws in areas such as investment and intellectual property protection, were adjusted to international standards to promote private investment.

During the trade liberalization of the two Latin American countries, the state-led import substitution policy was replaced by an export orientation as a new economic development strategy. The opening of the market facilitated access to foreign technology, such as plant varieties and agro-chemical inputs, and offered better export opportunities. These opportunities predominantly have a hemispheric and a regional dimension.

Closer hemispheric economic integration is pursued within the framework of the North American Free Trade Agreement (NAFTA), of which both Chile and Colombia wish to become members. NAFTA is of great importance to the expansion of Chile's exports, especially when taking into account the fact that the total value of the Chilean agricultural exports declined from US$ 3.3 billion in 1995 to 2.9 billion in 1997 (ODEPA, 1998). The country is a major exporter of fresh fruit, wine, cellulose, and wood, a large part of which is marketed in the USA and Canada. NAFTA is of equal importance to Colombian exporters of cut flowers, vegetables, and tropical fruit. Colombia is the world's second largest cut flower exporter. Around 80% of the cut flower export is sold in the USA (ASOCOLFLORES, 1994). Much is expected of the production of asparagus, mango and some tropical fruit juices. Tropical fruit is considered to be the Colombian export flagship of the next century (CCI, 1994).

Closer regional economic integration is the second major objective of the *apertura*. Colombia is a member of the Andes Pact (with Venezuela, Ecuador, Bolivia and Peru) and is supposed to enjoy a comparative advantage in agriculture in relation to the other Pact members. Colombia accounts for 60% of the farm output produced by the Andes Pact. Colombian potato and sugar cane production in particular has regional export potential (World Bank, 1996:16). The regional economic integration aspirations of Chile materialize within the framework of the *Mercado Común del Sur* (MERCOSUR), the free trade group of Argentina, Brazil, Paraguay, and Uruguay, of which Chile is an associate member. As regards the agricultural sector, MERCOSUR is assumed to entail fewer benefits than NAFTA (Quiroz, 1996:128). Public potato breeders, however, foresee a bright future for the export of both fresh and seed potato in the region (interview Lopez, 1996).

The economic liberalization in Colombia and Chile may be advantageous to agricultural exporters, but it has had a severe impact on producers of staple crops destined for the national market, such as rice, cotton, barley, and wheat. The lowering of protective measures allowed cheaper imported food to compete directly with nationally produced food, sometimes with devastating results for national farmers. Both in Chile and Colombia, the share of imports relative to national production of the main food crops has grown quickly in the past decade (Figures 6.2 and 6.3). In fact, the economic liberalization was a burden on an agricultural sector that was already under pressure. The position of the national producers has deteriorated due to periods of drought (in Colombia in 1992/93, and in Chile in 1996), and because of monetary measures such as a revaluation of the national currency, making national production even more expensive in relation to imports (Muchnik, 1996:122). Moreover, Colombian producers have to cope with dietary changes in their country, in which traditional food is gradually being replaced by industrially processed (Western) food. This development entails a decreasing demand for national crops, notably rice, while the demand for wheat and potato-based food is increasing (interview Moscardi, 1996). The expansion of domestic demand for products that can more cheaply be imported put enormous pressure on the national farming sector. Finally, the agricultural sector in Colombia is suffering badly from the rural violence caused by guerilla forces and cocaine producers.

Figure 6.2 Maize and wheat imports in percentage of national production in Chile (1990-1996)

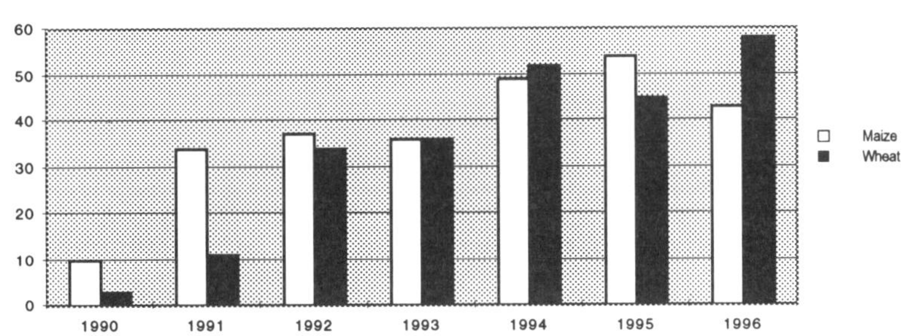

(Data derived from FAOSTAT, 1998)

Figure 6.3 Barley, maize and wheat imports in percentage of national production in Colombia (1990-1996)

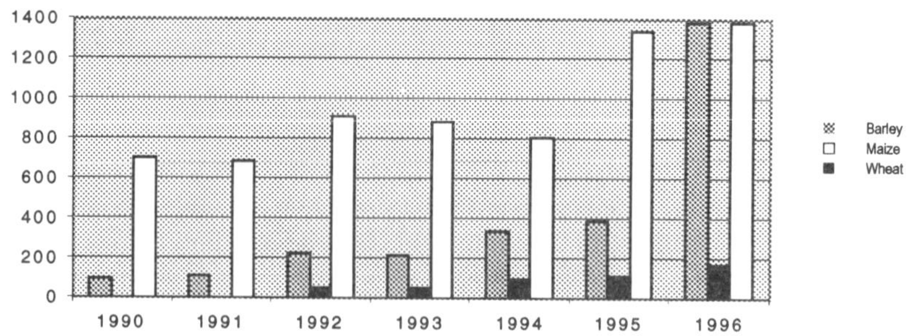

(Data derived from FAOSTAT, 1998)

The combination of economic liberalization with uncontrollable climatic, political and cultural circumstances has confronted the Chilean and Colombian governments with a major dilemma. In both countries only a handful of agricultural subsectors is equipped to compete on the international market. The majority of the rural population, which still constitutes a large part of the population[3], is by no means prepared to compete with farmers from, for example, Argentina or the OECD countries. Many farming families were already living in poverty before the economic liberalization set in, and their number is growing. In Colombia even one third of the rural population is found to be "extremely poor" (World Bank, 1996:5), and it is beyond doubt that the violence caused by the guerillas and the *narcotraficantes* is rooted in the hopeless situation in the countryside.

Nevertheless, a return to protectionist economic policies has in the past few years been considered neither by the succeeding Colombian governments, nor by the democratically elected governments in Chile. The policy to combat poverty in these countries has been a continuation of liberalization and of agricultural rationalization, although it is intended to give the market-led strategy a "human face" (Kay, 1996:4). The titles of some recent key publications on agricultural policy illustrate the prevailing intention that a strengthening of agricultural competitiveness is to be accompanied by alleviation of rural poverty (*cf.* Gonzales and Jaramillo, 1994; World Bank, 1995). Import tariffs up to 30% have been maintained for most crops in Colombia (World Bank, 1996:72), while the import of rice from Vietnam and Thailand has been prevented (interview Sanabria and Dabalos, 1996). In Chile, wheat, rice, maize, and oil crops have also enjoyed some form of protection and are excluded from hemispheric and regional trade agreements (Muchnik, 1996:117,119).

In both countries development programmes have been set up to equip small farmers better to conditions of the more open market. In the early 1990s, the Colombian government and private industry jointly designed crop diversification programmes to encourage small farmers to produce promising crops, such as vegetables, fruit and pulses, instead of wheat and barley, sisal, and cotton (interview Alvarado, 1996). The Colombian government terminated its support for the latter crops, on the grounds that Colombia has no competitive advantages in their production. The major consumers of wheat and barley, the Colombian millers and breweries, preferred to import these products, for reasons of price and quality (interviews Alvarado, 1996; Quintama and Pordnía, 1996). In another programme, the government attempted to improve access to land. In collaboration with the World Bank a programme was designed to help 70,000 landless farm families to purchase land by a 70% price subsidy. It is probably illustrative of the severity of the rural problems in Colombia that neither the crop diversification programme, nor that on land access, has been successful so far. Effective crop diversification is hindered, among other things, by the poor condition of the roads in Colombia which do not allow for adequate transport of perishables from remote rural areas to the large city markets (interview Isaacs, 1996). The land access was not functioning well, because only one hundred families were able to borrow the remaining 30% (in addition to the 70% subsidy) (Anonymous, 1996a).

In Chile, the democratic governments since 1990 have designed a distinct policy to strengthen the small-farmers sector, a policy referred to as the *reconversión*. It focuses on the increase in productivity of traditional crops, on improving processing and marketing, and on diversification away from traditional crops to forestry activities, livestock rearing or non-traditional export products, such as cut flowers (Kay, 1996:16-19). In principle the *reconversión* is aimed at all farmers, but a distinction is made between 'viable' and 'non-viable' farms. This is an important aspect of government intervention in agriculture and a well-tried policy in the USA and Europe. The Chilean government is actually limiting its support to farmers who have the potential to establish a farm that generates at least a family income. All other farm units, the *minifundistas*, are in fact ruled out by the conditions which farmers have to meet in order to receive technical assistance (Kay, 1996:14-20).

6.1.3 Conflicting agricultural production strategies in Chile and Colombia

When we compare the agricultural policies followed since 1990 in the two Latin American countries with the typology of production strategies presented above, we find a combination of elements from the market-led and the state-led strategies. In both countries, the government has actively pursued a further industrialization of the national agricultural sector. Although not being involved itself in the production of cut flowers, fresh fruit or wood, the government paves the way for private expansion in this area by concluding free trade agreements, and by adopting legislation for the improvement of intellectual property protection and investment opportunities.

Without any state intervention, the opening of the agricultural market would facilitate overall industrialization, considering that the import of cheap foodstuffs has a beneficial effect on wages. It would certainly also encourage a further agro-industrialization in other parts of the world. At the same time, however, it would probably wipe out most of the national farm sector in Chile and Colombia, which is not equipped to compete with the world's strongest, and often state-supported, farm systems. Some protective measures have therefore remained in both countries, albeit at a significantly lower level than before. Public policies for technology transfer, diversification, land access improvement and management can be considered as instruments of the government to speed up the industrialization of agriculture. The peasantry has to be converted into viable enterprises that can afford industrial inputs, sell on the market at competitive prices, and raise a family income at the same time.

During the interviews in Chile and Colombia we found support for both the market-led as well as the state-led strategies. Not surprisingly, the support for the state-led strategy was particularly voiced by the main Colombian producer *gremios*, such as the federations of rice and cereal producers and of agronomists, and the *Sociedad Nacional de Agricultura* (SNA) in Chile. Chilean fruit and wood exporters and the exporters of flowers, potato, sugar and vegetables in Colombia were among the main advocates of the market-led strategy, at least for the products they are involved in.

The Latin American exporters thus advocate a production strategy similar to that of the transnational crop development conglomerates, but with a different motivation. The conglomerates have a global approach. They deal with a great variety of basic food crops, which enable them to supply markets everywhere, while their diverse and multinational production base requires unhindered movement of capital and products around the globe. The slogan 'Feeding the World' reflects the increasingly global scope of thinking of the agro-industrial TNEs. The Latin American agro-exporters, however, have a different reason to advocate trade liberalism. Provided that they do not form a part of these TNEs themselves, the export producers are much more vulnerable to the volatile world market, because their production base is located in only one country and they lack a domestic market for their products. The Latin American exporters seek entry into the world market through agricultural niches and require a liberal trade strategy because both the technology and markets on which they entirely depend on are mainly situated in OECD countries.

The governmental agricultural policies in Chile and Colombia are directed towards further industrialization of agriculture. Also the attention to the poor rural population is based on the intention to incorporate the small farms into the agro-industrial system. However, this intention can only partly be effectuated. As indicated above, a large part of the poor rural population cannot enjoy the intended benefits of the government support programmes. As many farms are considered to be 'non-viable' production units, neither the government nor the established farm unions wish to invest in their development any longer. Their problems are considered to be a 'social' rather than a 'production' issue (interviews Jordán, 1996; Orella, 1996). Whereas in the main OECD countries the superfluous farm population could be absorbed by expanding industry or be supported by public social security schemes; in the two Latin American countries in question neither the industrial capacity nor social support could suffice to this end. This implies that a consi-derable part of the rural population is left to fend for itself. The needs of this group are primarily addressed by a wide variety of non-governmental organizations, including domestic organizations of *campesinos*, of indigenous peoples and of afro-americans; churches; nature-oriented, farm-oriented, and internationally operating NGOs; foreign aid donors; and also cocaine barons and guerrilla forces.

The common denominator of the NGOs we have interviewed is the concern for a further marginalization of a considerable part of the rural population as a result of agro-industrialization. They disapprove of the policies that prompt farmers to implement monocropping rather than diversification, because this would endanger food security. Striving for competitiveness requires the use of commercial plant varieties and associated agro-chemicals, which raise production costs and worsen contamination problems. The organizations therefore aim at an autonomous production strategy, i.e. production independent from the international market or even from the market economy as such. Their goals include food security for resource-poor, rural communities, and a reduction of production costs through more efficient use of locally available natural resources.

The Colombian and Chilean nature and farm-oriented NGOs generally maintain intensive contact with like-minded NGOs and donor organizations in OECD countries. The conceptual or ideological similarity between these organizations is high, and enables transnational coalitions between Northern and Southern organizations in the promotion of what we have referred to as the 'non-industrial, farm-oriented production strategy'. However, the Colombian and Chilean position in the international division of labour in agriculture is different from that of the home countries of the internationally operating NGOs. This difference seems to induce South-North tensions in the interpretation of the non-industrial production strategy. Whereas Northern advocates of this strategy to a large extent have ideological or romanticist grounds, their Southern counterparts tend to have more pragmatic considerations. The misery that results from their marginal position in an industrializing agricultural sector leaves the 'non-viable' farmers no other option than to rely on locally available resources and traditional knowledge. Hence, in Chile and Colombia, conservation of biological diversity and the use of landraces is not an objective *per se,* but a necessary condition for sustaining the resources required for achieving a minimum of food security.

6.1.4 The PGR conflict as a 'spark' between rival agricultural production strategies

In a first step to explain the controversy on crop development in the Third Agro-Food Order we have introduced three ideal-types of agricultural production strategies that can be distinguished in the market economy. This typology of rival strategies applies to OECD and developing countries alike. However, the political choices for either of the strategies have more severe social implications in developing countries, as agriculture still constitutes a major pillar of their economy.

The neoliberal strategies pursued in Chile and Colombia have revealed a split in the agricultural sectors of both countries. As soon as the protective wall was lowered, it became clear which crops could be produced competitively, i.e. which could compete with imports or be exported, and which could not. The competitiveness of crops also indicated which producers could win from liberalization and which would lose. Producers of export crops tend to benefit from a market-led agricultural production strategy, while producers of the main staple crops require and demand governmental protection.

The basic assumption of the neoliberal policy is that an open trade regime leads to higher growth rates, which will create employment and so contribute to poverty reduction. This may happen in the long term; meanwhile many Chilean and Colombian farmers cannot participate in liberal industrialization process and are further marginalized. They have to rely on their own improvements in production, and support from NGOs and foreign donors.

In the following sections we explain how the growing gap between industrialized and non-industrialized agricultural production strategies in Chile and Colombia helps an understanding of the controversy over crop development. The tension between production strategies is not restricted to the question of which farmers are able to survive neoliberal policies. Each production strategy includes a particular view on how conservation and IPR protection should support agriculture. The controversy on conservation and IPR protection is, in our view, therefore caused by controversy on production strategies.

6.2 The state loses control over national conservation

In the middle of the Chilean desert lies a heavy bunker, the design of which closely resembles that of a ballistic missile launch installation. But instead of rockets, it contains a 'core' collection of Chile's national biological diversity. The fact that this bunker was mainly financed with Japanese funds illustrates the general development that is at issue in this section, namely that the conservation of seed, plants and plant material is no longer dominated by national agricultural interests only, but also serves the interests of the transnational crop development conglomerates.

The international involvement in national and local conservation strategies has caused considerable opposition in Chile and Colombia. The fact that most opposition is brought to bear by small farmers' groups and indigenous people's organizations leads to the second issue of this section, i.e. the PGR conflict as an element or 'spark' in the friction between non-industrialized and industrialized production strategies. What role does the PGR conflict play in the survival of marginalized groups? What do these groups expect from the implementation of the CBD?

6.2.1 The disconnection of national agriculture and conservation

One of the conclusions we were able to draw from interviews with breeders and agricultural producers in Chile and Colombia was that a uniform pattern in conservation does not exist. Different market orientations result in different strategies in breeding and consequently in conservation. Opinions about the importance of conservation of landraces and wild relatives even vary per crop. The background of this disparity is the new market-orientation in agriculture in the two countries. The necessity to produce against internationally competitive prices makes producers rely increasingly on imported, high yielding varieties bred by the transnational crop development industry. The result of this tendency seems to be a situation in which the focus of the national conservation efforts is shifting towards the interests of the transnational crop development industry.

It is common knowledge among plant breeders that no country is self-sufficient when it comes to plant genetic information. Agricultural production in all countries is based on the historical, worldwide exchange of plants and seeds that have been used in local breeding programmes. Examples are the wheat and rice production in Chile and Colombia. Apart from foreign varieties, locally bred plant varieties contain plant material from various sources, including varieties of the CGIAR, foreign commercial varieties, national varieties, and locally collected landraces (interviews Accatino, 1994; Von Baer, 1994 and 1996; Muñoz, 1994).

During the Third Agro-Food Order a qualitative change can be detected in which developing countries increasingly import finished varieties from the crop development conglomerates and independent breeding organizations in OECD countries. Since the import of foreign varieties makes domestic breeding efforts superfluous, the direct importance of locally available resources of plant genetic information for national agriculture tends to decline. This increasing incompatibility of conservation and national agriculture can be illustrated with the example of maize conservation.

In Chile, a considerable variety of maize landraces can be found. Maize was introduced into Chile by the Incas from Peru, 200 years before the Spaniards arrived. Ever since, maize has been adapted to Chile's many different agro-ecological conditions. The resulting genetic diversity induced the CGIAR institute IBPGR (now IPGRI) to co-finance a maize collection programme, which was carried out by the *Instituto Nacional de Investigación Agropecuaria* (INIA) during the 1980s. While duplicates of the collected material were transported to the U.S. National Seed Storage Laboratory (NSSL) (INIA, 1990:5), the collection seems hardly relevant for maize production in Chile itself. Around 90% of the maize acreage is planted to hybrid varieties imported from American and European enterprises. Only 10% of the land on which maize is produced is used to grow *choclero*, a traditional open-pollinated maize which is used to prepare *humita*, a national maize pasta. One of the seed firms we interviewed, Agro-Tuniche, showed an interest in this market segment, but only as a hobby. The firm is the local maize seed dealer of the U.S. company Mycogen. Its Chilean vice-president used the INIA maize catalogue for the collection of landraces in Chile and Peru, in order to set up a small breeding programme. From a commercial point of view, however, the variety is of marginal importance (interview Alamos, 1996).

A similar situation regarding maize can be found in Colombia. The country hosts a wealth of genetic diversity in maize. A third of the accessions of the national seedbank maintained by the *Corporación Colombiana de Investigación Agropecuario* (CORPOICA) consists of maize seeds. The Colombian maize collection is playing an important role in the international LAMP collection network, which is supported by the USDA and Pioneer Hi-Bred (see section 4.5.2). The project also facilitates a free transfer of Colombian maize seeds to the USA. The function of the maize collection for Colombian agriculture is tending to decline, however. CORPOICA runs a maize-breeding programme, but this primarily involves non-hybrid varieties, which are not adjusted to the demands of the processing industry. Even though most of Colombia's maize acreage is still planted to the local open-pollinated varieties, the market for American hybrid varieties is growing (interviews Barbosa, 1994; Copete, 1994; Torres, 1996; Urres, 1996).

The reliance on imported varieties is particularly strong in the case of non-traditional exports. Even though in some cases there is abundant local availability of genetic resources, there are no national breeding programmes in support of the new export orientation. The few initiatives to set up a breeding programme based on endemic plant material could not attract any interest from the private industry, as the following examples show.

In Chile, thanks to favourable agro-ecological conditions and short distances to the major harbours, export production of strawberries is a profitable business. The strawberry is native to Chile, and the world's main commercial varieties, which have been bred in the OECD countries, are partly based on species that were collected along Chile's vast coastline. Domestic producers, however, are not interested in the native resources. The local landraces are good in taste and size, but their yields are low. In the early 1990s, the Catholic University of Chile initiated a strawberry-breeding programme with foreign financial assistance, but the programme was terminated as soon as the foreign donations dried up. Chilean export producers prefer to obtain their varieties from the USA and Europe, rather than invest in their own breeding programme (interview Legarraga, 1994).

A similar attitude can be found among wood exporters. Around 40% of the value of Chile's agricultural exports consist of wood products, but this prominent economic interest is not translated into tree breeding or conservation programmes. Various parts of southern Chile are still covered by primeval forest. Since around 60 indigenous wood species are threatened with extinction, various universities have started to collect shoots and other material of the trees for conservation and propagation. There is no industrial interest in the universities' genebanks, however. The industry concentrates on fast growing species, including *Pinus*, *Eucalyptus*, and *Populus*, which are imported from Australia, the USA and Europe. The native trees grow too slowly for commercial exploitation (interviews Benoit 1996; Lara, 1996; Seemann 1996).

We have mentioned earlier that Colombia is the world's second largest cut flower exporter. Most production is based on sub-tropical varieties produced in OECD countries, even though the country's tropical forests contain a wealth of wild flowers. The export producers have no commercial interest in these resources. This was the reason why they turned down an offer to collaborate with researchers of the University of Antioquia in Medellín, who set up a research and breeding programme on the native *Heliconia* flower. Of the 250 existing *Heliconia* species, 92 can be found in Colombia, of which about 48 are endemic. As American researchers are also working on a breeding programme for this flower, the University is now looking for Asian investors and technology suppliers, in order to breed a commercial Colombian variety (interview Arthehortúa, 1996).

Another Colombian example is in fruit production. In Colombia, there is optimism about the export potential of tropical fruit species that have not yet been marketed. The *Corporación Colombia International* (CCI) was specifically established by the government and businessmen to explore U.S. and European niche-markets for Colombian fruit and vegetables. Referring to New Zealand's success in exporting kiwi fruit, some Colombian businessmen even consider the *pitahaya* or *guanábana* to be the Colombian 'kiwi' of 2010 (Anonymous, 1995a:25). The CCI, however, does not intend to encourage domestic breeding or conservation programmes of locally available plant species (*cf.* CCI, 1994; interview Isaacs, 1996). Initial exploration of Colombia's richness in tropical fruit was only carried out by two public international research institutes, CIAT and IPGRI (interviews Debouck, 1996; Williams, 1996).

The lack of interest in breeding on the part of export producers in Chile and Colombia seems to have two underlying causes. The first is that the Latin American producers consider themselves to be technically behind. It takes a very long time and considerable capital investment before commercial varieties can be developed from wild plant species. Relying on foreign varieties bred in OECD countries is much cheaper (interviews Baldrich, 1994; Camacho, 1994; Alamos, 1996; Concha, 1996). This pragmatic attitude is illustrated by a statement of the Director of *Viveros Requinoa,* a Chilean fruit producers' association: "Excellent material can be bought from foreign breeders in California or in France. They have sometimes 40-50 years of experience. We will not be able to perform better, not even through large investments in national research programmes. Why compete when we can buy it?" (interview Concha, 1996).

In addition to the technical capacity, it is generally conceived in the two countries that the absence of investment in breeding is grounded in the tradition of political instability in Latin America. Domestic private companies are not inclined to develop long-term investment strategies. An official of *Fundación Chile*, an organization set up to encourage private entrepreneurship in Chile, puts it as follows: "All older people know that the policy of today may be different tomorrow. Even though there has been considerable stability in the past 15 years, still there is no trust that this period will last for a longer time." (interview Lorenz, 1996).

The choice of producers of non-traditional export commodities to import plant varieties from the crop development industry in OECD countries, rather than breeding new varieties themselves, is not only strengthening the international division of labour in crop development, but it also has consequences for the interest in conservation. Where domestic breeding programmes are absent, there is little national incentive for the collection and conservation of landraces and wild relatives. If these materials are being collected, it is basically because of funds made available by international or foreign organizations. The consequence is that the objectives of conservation are primarily directed towards the requirements of the foreign crop development industry, while the direct relevance for national agriculture is unclear.

This situation is illustrated by the new INIA seedbank, which is equipped for long-term conservation. Although it stores Chile's national collection, it was the Japanese Agency for International Cooperation (JICA) that provided most funds. JICA built the Chilean seedbank and helped to fill it. In the period 1988-1994, JICA and INIA jointly organized eight expeditions to various regions in the country in order to collect forage, strawberry, and tomato species (interview Cubillos and Lobos, 1996). The large Japanese involvement is not merely an act of charity. Japan and Chile cover similar climatic zones, albeit on opposite sides of the equator. Chile's biodiversity may thus help to secure Japan's long-term access to genetic information in support of its national, highly industrialized agricultural production system.

We could only find a few examples of increased international competitiveness coinciding with an interest in local conservation; Colombian potato breeding is one of them. Due to climatic differences, the high yielding potato varieties from Europe or Canada cannot grow in the Andes. The absence of superior foreign varieties, and the good economic prospects for the regional export of local potato landraces (*criollo* varieties), generates an interest in the nationally available potato resources for national breeding programmes (interview Arevalo and Lugán, 1996).

Also in Colombia's coffee breeding programme, there is clear link between national collection and production. The export of coffee is of major importance to Colombia's economy. In 1993, coffee exports amounted to nearly one quarter of the value of Colombia's total exports (World Bank, 1996). The coffee producers, organized in the *Federación Nacional de Cafeteros* (FEDECAFÉ), possess one of the most valuable coffee collections and breeding programmes in the world (interview Herron, 1996).

In sum, a uniform pattern in conservation does not exist in Chile and Colombia. Only one general trend can be discerned in which national conservation and crop development are becoming less compatible, due to the market-orientation in agriculture. In various important staple crops, high-yielding foreign varieties are increasingly preferred over domestically bred varieties. For non-traditional export commodities, the breeding of the country's own varieties is not considered to be cost–effective. Even for crops of which genetically diverse plant material is locally available, there is no private investment in breeding, except for a few cases in which suitable foreign varieties fail. The preference for imported varieties depresses national breeding efforts and diminishes the value of native species as a breeding base for national agriculture. This development has not terminated investments in national conservation, however. On the contrary, the conservation of native species secures the long-term access to genetic information of the transnational crop development industry. Whereas the increased market-orientation in agriculture tends to disconnect conservation and crop development at national level, a new connection between the national conservation efforts in individual countries and plant breeding seems to be established at transnational level.

6.2.2 The national controversy over biodiversity conservation

The ratification of the CBD in 1992 codified and confirmed the right of countries to subject diverse biological organisms within their territories to national sovereignty. The CBD sets out the general rules for the conservation and access to biodiversity, but the precise implementation of these rules is left to the signatory countries themselves. This has caused considerable political controversy in Chile and Colombia. On the one hand, the market-orientation in agriculture in both countries has made many actors cautious of an orthodox interpretation of the CBD. Hindering foreign organizations in exploring national biodiversity may backfire when national organizations require access to genetic resources abroad. On the other hand, marginalized farmers' and indigenous peoples' organizations have seized the national implementation of the CBD as an opportunity to improve their socio-economic position. They claim communal rights to locally available biodiversity in an effort to enforce their struggle for more autonomy.

In the previous section we concluded that the market orientation in agriculture in Chile and Colombia tends to result in a greater reliance on foreign high-yielding plant varieties. Producers therefore have an interest in shaping the conditions that facilitate the import of foreign varieties. One measure is to adopt IPR protection legislation, the other is to keep access to the national flora relatively open. The word 'relatively' must be emphasized in this respect, because also several researchers and export producers in Chile and Colombia had mixed feelings about foreign organizations collecting native plant species in their country for the development of protected, commercial varieties (interview Patiño, 1994; Arevalo and Luján, 1996; Arhehortúa, 1996; Madrid, 1996). They mentioned several examples of native species that are exploited by the foreign crop development industry, including strawberry and tree tomato (fruit), *quinoa* (grain), *Alstroemeria* and *Heliconia* (flowers), and potato. The uneasiness, however, seemed largely to be a self-criticism, in that domestic organizations were either not interested or not able to invest in the commercial exploitation of local biodiversity themselves.

A prevalent perception among researchers and producers was that ready access to foreign plant material was much more important than preventing the exploitation of native species by foreign organizations. The head of INIA's genetic resources programme in Chile argues that "as most other countries, we rely almost exclusively on foreign germplasm, so we should cherish possibilities for exchange.

We therefore cannot permit ourselves to maintain a restrictive access policy" (interview Cubillos and Lobos, 1996). For similar reasons the Colombian national PGR coordinator at CORPOICA supports open access: "Like the USA we depend on the inflow of foreign material to keep our agricultural sector going. The CBD could sincerely diminish the exchange of economically important material." (interview Lobo, 1996). The concern that any hindrance of access to the national biodiversity would backfire on Chile's access to foreign propagating material of the *Eucalyptus* and *Pinus* species was even the reason for an attempt by several members of the Chilean Congress to exclude tree species from the implementation of the CBD (interview Muñoz and Flores, 1996).

CGIAR centres, CIAT and IPGRI Americas, in Colombia were very concerned about the negative impact of the Colombian CBD regulation. Breeders and curators of both institutes felt that the CBD might harm the free access policies on which the CGIAR network has built its collections in the past three decades. CIAT's conservation strategy will be most directly affected by the Andean Pact Decision 391 of the Common Arrangement on the Access to Genetic Resources, which arranged the implementation of the CBD (Comisión del Acuerdo de Cartagena, 1996). Decision 391 specifies sovereign and indigenous rights over PGR, and allows signatory countries to replenish established access policies - such as CIAT's free access policy (interview Roca, 1996). Since 1992, CIAT seedbank curators have increasingly been confronted with conditional requirements for access (such as reciprocity) or total refusals of requests for landraces and planting material. IPGRI Americas' director even fears a "black market" among breeders that want to circumvent the increasingly complex national access regulations (interview Williams, 1996).

Whereas Chilean and Colombian breeders and producers have little interest in a strict access regulation, the many small farmers' and indigenous people's organizations tend to have an entirely different view on this issue. Protest against the free access regime and the unequal benefits derived from PGR has even become one the spearheads of their campaigns to improve their living conditions.

The PGR conflict first emerged during the drafting process of the FAO Undertaking on Plant Genetic Resources in 1983 (see section 1.1). Although the FAO debates lasted throughout the 1980s, they were confined to governmental representatives and northern farm-oriented NGOs. During the debates, PGR was merely treated in terms of their economic value for the northern agro-industry. The implementation of the CBD anchored the PGR conflict in the much broader discussion on the sustained use of biodiversity. In these discussions, small farmers' groups and indigenous people's organizations were not only recognized as victims of the gene-drain, but also as the 'custodians' of the earth's biodiversity. This recognition has a considerable emancipating effect on communities that still rely on non-industrialized forms of production. Most is expected from CBD Art. 1, which stipulates a "fair and equitable sharing of benefits arising out of the utilization of genetic resources"; Art. 8(j), which recognizes the importance of "indigenous and local communities embodying traditional lifestyles relevant for the conservation and sustainable use of biological diversity"; and Art. 15(d), which demands "prior informed consent" of the owners or custodians of PGR (UNEP, 1992).

The intensity of the PGR conflict in Chile and Colombia suggests that it cannot be considered as an isolated political issue. Questions on access to PGR provoke most controversy when they interfere with broader, long-standing political controversies, particularly territorial and ethnic rights. The PGR conflict is often further intensified when local pressure groups are able to strike transnational coalitions with other organizations. The introduction of cheap and fast communication lines, particularly the fax and more recently the Internet, have given Chilean and Colombian pressure groups access to a growing amount of information, finance, and political and moral support from like-minded organizations worldwide. Both the intensity and global extent of the attention local PGR issues can attract at short notice have made it increasingly difficult for the Chilean and Colombian governments to neglect the political demands of marginalized groups.

In Colombia, matters of access and exploitation of native PGR result in conflicts between the government and indigenous people's organizations only when they are directly connected with broader issues of territorial and ethnic rights. The Colombian constitution recognizes territorial rights of indigenous groups, as well as the economic and cultural value of biodiversity within their territories[5]. This has allowed them to build up considerable political pressure on the national government on biodiversity access questions. The implementation process of the CBD in Colombia has even led to a relatively successful coalition between indigenous people's organizations and nature-oriented NGOs. With the support of several congress members, the coalition succeeded in establishing an Ad Hoc Working Group on Biodiversity at the Ministry of Agriculture. Protected by its official status, the Working Group drafted Colombia's national biodiversity conservation strategy. In 1994, the Group was able to gain regional and international support for its actions during the drafting process of Decision 391 of the Andes Pact (interview Pombo, 1996). However, the Group encountered serious governmental opposition, especially after it had attempted to prevent the entrance of oil-drilling and wood logging companies into indigenous territories on the grounds of territorial and cultural rights of the indigenous peoples. The Colombian government succeeded in outmanoeuvring the Group from the national political arena in September 1994, when the new Ministry of Environment was founded, which took over the implementation of the National Biodiversity Strategy from the Ministry of Agriculture. From then on, the status of the Group diminished to that of an advisory panel (interviews Flórez, 1996; Fonseca, 1996; Pombo, 1996).

Mapuche indigenous communities in southern Chile have also attempted to link PGR conflicts to the long-term, structural problems - problems notably of territorial and ethnic rights - that they have faced since colonization by the Spanish. Although these problems gained limited international attention, the exploitation of PGR of potato landraces of the Mapuche by foreign breeders was rapidly picked up by many international nature and farm oriented NGOs (interview Herrera, 1996). The international criticism came to a climax in 1995, when it was discovered that

Mapuche blood and hair samples were taken for analysis in the U.S. Human Genome Diversity Project (HGDP) without what in CBD wording is called "prior informed consent". In several poignant RAFI press-releases the HGDP was dubbed a "vampire project" (RAFI, 1995b). Although Mapuche representatives welcome the large international attention for the potato and blood cases, they are concerned that the PGR access question will not improve the fate of the Mapuche in Chile, if it is not placed within a broader territorial, ethnic, and socio-economic context (interviews Heunchulaf, 1996; Nain and Santander, 1996; Melillan and Antilef, 1996). This concern was clearly summarized by the director of the Chilean National Corporation for Indigenous Development (CONADI): "Of course we don't want to be considered as a bag of genes. But there are many more urgent problems which hit us almost every day: poverty, inequality, lack of legal status, discrimination, racism ..." (interview Heunchulaf, 1996).

The critique of the Mapuche in Chile and the difficulties of the Ad Hoc Working Group on Biodiversity in Colombia indicate that the real political significance of the PGR conflict lies in its relation to the socio-economic position of marginalized groups. Although the interests of small farmers and indigenous communities are often addressed in international biodiversity conferences by transnational NGOs, the FAO, UNEP, World Bank, or the Conference of Parties of the CBD, the communities themselves are often unable to translate the outcome into political measures that improve their socio-economic position. This experience has raised considerable scepticism among the organizations of farming and indigenous communities about the real purpose of the CBD. The difficulties they face when they attempt to connect PGR issues to 'hard' political issues has led to the belief that the CBD is more in the interest of OECD countries.

The inability of marginalized groups really to benefit from the juridical framework offered by the CBD is the bottom line of argument of the Latin American Institute for Alternative Legal Services (ILSA, Colombia). Flórez of ILSA comments that although the CBD recognizes the role of indigenous groups in the conservation of biodiversity, they are not in a position to enforce a mechanism of "fair and equitable sharing of benefits" (Art. 1) that really serves their interest. An additional constraint is that the heterogeneity among the communities is likely to result in fragmented policies on benefit sharing (interview Flórez, 1996). The Centre for Technology and Education (CET, Chile) holds that international NGOs have created a climate in Chile in which indigenous people's organizations have been forced to negotiate complex access regulations. Many small groups lack the institutional and financial capacity to negotiate their interests adequately (interview Montecinos, 1996).

The limited capacity to fight political battles over PGR has made farmers' groups and indigenous people's organizations dependent on their alliances with OECD-based NGOs which do possess sufficient financial and human resources. The fact that especially nature-oriented NGOs, which tend to consider indigenous groups to be custodians of biodiversity, have often taken the initiative in helping developing countries to implement the CBD, has also caused distrust. A. Green, the main representative of the National Organization of Indigenous Communities

(ONIC, Colombia) warns that too explicit attention for biodiversity as 'scarce nature' may also backfire on their interests as active users. He cites the establishment of the Katio nature park in the Chocó region, which implied the expulsion of indigenous Embena and Wandana communities (interview Green, 1996). A similar concern is expressed by the Colombian Ad Hoc Working Group. In 1994 the Group had to put much effort into diminishing IUCN's influence in the drafting process of 'Decision 391' of the Cartagena Agreement. The Group disagreed with IUCN's strict economic interpretation of the CBD access regulation. During a preparatory meeting early in 1994, IUCN had proposed a compensatory mechanism resembling the 'bioprospecting' system applied by pharmaceutical conglomerates (see section 5.2.2). This interpretation, in combination with IUCN's limited consideration of ethical, social and agricultural issues brought up by the indigenous people's organizations, created an atmosphere in which the Group felt overruled by a northern-based NGO (interview Pombo, 1996).

In sum, for small farmers and indigenous peoples the conservation and use of biodiversity is just one element of an age-old, non-industrialized agricultural production strategy. Battles over the genetic resources this biodiversity contains can therefore also be considered as one element of a much larger struggle for the survival of the non-industrialized production strategy. The recent attention for biodiversity allows marginalized groups to link PGR conflicts with issues that may ease their social and economic position, such as territorial or ethnic rights. Simultaneously, the PGR conflict offers opportunities to attract international support for their position. In Chile and Colombia this support often does not meet the most urgent demands of the small farmers' groups and indigenous people's organizations. The rationale of northern environmental NGOs for helping these groups is mainly based on a mixture of concern for pressures on both farming communities and biological diversity, while it is not always clear which of the two is considered to be the means for sustaining the other.

6.2.3 National conservation for the transnational crop development industry

While controversies on conservation are gradually entering the national political realm, it seems as if the Ministries of Agriculture in Chile and Colombia are not the prime actors any more in the design of conservation strategies in both countries. Instead, the national conservation strategies are increasingly embedded in an international division of labour in crop development in which international, rather than national agricultural interests prevail. We have discussed three tendencies that contributed to this development: (a) the market orientation in agriculture, (b) the growing importance of foreign donors in national conservation; and (c) the growing impact of the CBD on national conservation strategies. The question that now arises is what is the driving force behind this development?

One part of the answer lies in the growing interest of OECD countries in sustaining global conservation. To continue to be able to develop new varieties,

OECD-based conglomerates require a permanent access to an increasingly broad 'genepool', the greater part of which is still available in developing countries. Chile, and especially Colombia, possess a considerable share of that genepool and have therefore become the beneficiaries of GEF programmes, bilateral aid programmes, and corporate funds to maintain and extend national collections. The other part of the answer lies in the growing dependence on high-yielding, foreign plant varieties. In view of the opening of the agricultural markets in Chile and Colombia, the usage of such varieties is required if national production is to be internationally competitive. The direct interest in the national conservation projects with foreign finance thus lies with the crop development conglomerates and independent breeding organizations in OECD countries. Domestic producers may benefit indirectly when they import the foreign, protected plant varieties.

The interest of the transnational crop development conglomerates in maintaining the global genepool, and the access to it, sheds new light on the interest of OECD countries in the implementation of the CBD. The CBD enables signatory countries to restrict the access to PGR. It seems unlikely, however, that either the Chilean or the Colombian government will use the CBD in this way. The ready access to foreign plant varieties and germplasm, which is required for the market-led agricultural production strategy, may come in danger when the door to the national biodiversity is closed. Another explanation lies in the decrease in public funding for national conservation, inducing seedbanks to accept finance - and conditions - from foreign and international organizations. As a result, the CBD seems to facilitate both conservation of and access to diverse biological organisms. Most developing countries demand some form of "primary informed consent", or a symbolic compensation for the export of landraces and wild relatives, e.g. payment for shipment costs. The CBD, however, has facilitated the formulation of such regulation on the basis of internationally agreed rules and principles. They may involve some transaction costs, but the overall benefit seems to be for the transnational crop development conglomerates. The CBD creates worldwide clarity on what genetic information is available, where it can be found, and under what conditions it is accessible. From this point of view, the CBD contributes to the advancement of the market-led, export-oriented production strategy that prevails during the Third Agro-Food Order.

Although the implementation of the CBD still depends on decisions of national governments, the conditions for such decisions are increasingly set by transnational actors. Particularly internationally operating nature-oriented NGOs, notably the IUCN, WRI, and WWF have become very active in designing 'global' conservation strategies that are actively promoted in developing countries that implement the CBD. Although these organizations do recognize the rights and problems of the marginalized groups, they intend that both non-industrialized *and* industrialized agriculture should benefit from biodiversity conservation. This non-discriminative stance denies the rivalry between the two strategies. Non-industrialized farming and indigenous communities receive recognition and sometimes funding for their role as custodians of biodiversity, but this support is hardly ever framed in broader programmes geared to general socio-economic improvement. OECD countries, on the other hand, do benefit from the genetic information that the CBD related global conservation programmes generate. This fact makes the current global conservation strategies less politically 'neutral' than they are often presented to be.

The gradual integration of Chilean and Colombian collections and uncollected native species into the global 'gene-pool' has revived protests from those groups depending on a non-industrial, farm-oriented production strategy. They consider the free global exchange of locally used plant material to be another, unwanted aspect of an increasingly industrialized, technology driven agricultural production strategy. The recognition in the CBD of the rights of farmers and indigenous groups as custodians of biodiversity has allowed these groups to generate considerable attention for their marginalized position. The real political gain of the CBD will, however, depend on their ability to link the PGR conflict to other, long-standing ethical and territorial questions. In Chile and Colombia, this seems rather difficult because the main thrust of governmental policies during the implementation of the CBD is to prevent the PGR conflict from 'spilling over' into other political issues.

Summary

This section has illustrated how the market orientation of Chilean and Colombian agricultural producers has enlarged their dependence on foreign plant varieties. This development has a somewhat contradictory impact on the use of national collections. On the one hand, the import of finished varieties is likely to diminish the national breeding capacity and the active use of national collections by breeders. At the same time, however, both Chilean and Colombian collections are becoming part of internationally financed conservation programmes which seem to serve the long-term interests of the transnational crop development conglomerates.

This tendency, in which private, transnational interests increasingly govern national conservation criteria in Chile and Colombia, is opposed by domestic farm and nature-oriented NGOs and indigenous people's organizations. In an effort to improve the socio-economic position of marginalized farm communities, they try to use the national implementation of the CBD to reinforce their long-standing territorial, socio-economic, and ethnic rights.

6.3 Controversy over the intellectual property protection of biological organisms

Intellectual property protection of plants has probably been the most disputed issue of crop development in the past decades. IPR protection is expected to encourage private organizations to invest in plant breeding research so that new varieties are developed to the benefit of the agricultural sector. However, many organizations in OECD and developing countries strongly oppose such protection for economic, ethical, or cultural reasons. In this section we show that the reason for being in favour of or against IPR protection of plant-related innovations corresponds with the preference for one of the three agricultural production strategies that we have categorized above.

Three different agricultural producer interests can be distinguished: (a) the agricultural exporters, (b) the breeders of national food crops, and (c) the peasantry. We indicate that the opposition to IPR protection in Chile and Colombia comes entirely from organizations that represent or support resource-poor farming communities and indigenous peoples, as well as from nature conservation NGOs.

6.3.1 IPR and plant breeding for export crops

In the previous chapter it was explained how in the 1990s the principles of legal plant protection were exported from OECD countries to the rest of the World. They were accepted by most countries as part of the process of economic liberalization they were undergoing, and were formalized as a new international standard through the adoption of the TRIPS agreement in 1994. Both Chile and Colombia had anticipated this new IPR agreement. They enacted PVP legislation in the mid-1990s, acceded to UPOV in 1996, and also allowed the granting of patents on plant genetic material.[6] The protection of intellectual property rights of the foreign plant breeding industry fitted well into the neo-liberal economic policy followed in Chile and Colombia. Therefore, the advocates of IPR protection of plants in these two countries are primarily those organizations that benefit from the economic liberalization: the producers of export commodities.

As we have pointed out in previous sections, plant breeding in fruit, vegetables, cut flowers and trees is almost non-existent in both Chile and Colombia. The producers of these non-traditional export commodities are nevertheless in favour of IPR, not because it stimulates them to invest in plant breeding, but because it facilitates the import of foreign plant varieties. Indeed, around 45% of all PVP titles in Chile were granted on foreign fruit and flower varieties until 1996 (interview Messina, 1998) while (as already mentioned), the first wave of PVP applications in Colombia in 1997 mainly involved foreign cut flower varieties (UPOV, 1998:9). In the period before the PVP system was in place, the Colombian and Chilean producers also had access to foreign varieties, but this was based on gentlemen's agreements and contract law. Foreign flower breeders even used to organize annual fairs in Colombia to show their varieties to export producers. However, the absence of plant-related IPR protection had a price. It is generally considered by the export producers that the foreign breeders offered neither their latest varieties nor the best germplasm quality, and that the royalties were higher and the licensing conditions less favourable (interviews Camacho, 1994; Patiño, 1994; Arango, 1996; Concha, 1996).

In Colombia, the few producer organizations that have their own breeding programmes, such as FEDECAFÉ (coffee), CENICAÑA (sugar cane), and FEDEAR-

ROZ (rice), may enjoy a genuine benefit from IPR. Like the transnational crop-development conglomerates, the world's main advocates of plant-related IPR, the Colombian breeding organizations have a technological edge to defend, though at a regional rather than a worldwide level. They have a particular interest in protecting their germplasm from exploitation in competing countries in the Andes Pact, and in Central American countries. Particularly in Ecuador, Venezuela and Guatemala, Colombian varieties are often propagated and used by farmers (interviews Cock, 1994; Herron 1996; Sanabria and Dabalos, 1996). Protection of this competitive advantage of Colombian agriculture vis-à-vis neighbouring countries was a reason that the Colombian Ministry of Agriculture promoted the adoption of a PVP system within the framework of the Andes Pact (interview Rueda, 1996).

6.3.2 IPR and plant breeding for national market commodities

The second group of advocates of plant-related IPR consists of breeding organizations that have plant breeding programmes dedicated to national market crops. The adoption of PVP legislation was especially welcomed by the small group of, in total 10 to 15, domestic plant breeding organizations in Chile and Colombia: private firms, producer organizations, and public institutes. The agricultural crisis in both countries in the first part of the 1990s had a considerable impact on these organizations. The crisis reduced the area of land under cultivation, and induced farmers to use on-farm saved and informally traded seed, rather than commercial seed. The resulting contraction of the seed market had a negative effect on the profit margins of private plant breeders. PVP was therefore welcomed as a means to protect or expand their share of the seed market.

The public agricultural research institutes in the two countries had a similar reason to promote PVP. INIA in Chile and ICA/CORPOICA in Colombia have both been subject to considerable budget reductions following the embracing of neo-liberal economic policies in their respective countries. Protection of their new varieties offers the institutes the opportunity to tap additional income by collecting royalties. Seed sales are important for the institutes. About a third of INIA's income was accounted for by seed sales in the mid-1990s (interviews Portilla, 1994; 1996).

The domestic plant breeding organizations have advocated PVP rather than patents for genetic information, because they generate new plant varieties, not patentable inventions. Moreover, due to a lack of financial and technological capacity, they breed open-pollinated rather than hybrid varieties. This means that domestic breeding organizations cannot rely on built-in protection against unauthorized propagation and therefore have an interest in legal plant variety protection.

The degree to which unauthorized propagation occurs differs considerably among crops, due to climatic, biological and technical factors. In Chile, it is considered to be a problem predominantly by the wheat breeders. While there is considerable authorized on-farm seed saving of wheat, around 40% of the wheat seed supply reportedly consists of seed produced without authorization (interview

Portilla, 1996). The main target of PVP for domestic Colombian breeders is the unauthorized propagation of especially soyabean, rice, and non-hybrid maize varieties. It has been estimated that around 20% of the seed demand of these crops is produced and sold without the authorization of the breeder (interview Villota, 1996).

In order to exercise their rights collectively, the Chilean plant breeding organizations, in 1995, set up a Committee of Breeders of New Plant Varieties, as a part of the *Asociación Nacional de Productores de Semillas* (ANPROS). The Committee's president is provided by the sole private wheat breeding firm in Chile, while INIA and the fruit export producers' association provide the vice presidents. The committee is working vigorously on the detection of unauthorized propagation of plant varieties and the enforcement of PVP. Because neither the farmers, seed producers, nor the judiciary is familiar with the PVP system, the committee has hired journalists to publish on PVP occasionally, and it has also designed and financed a specific training programme for lawyers. At the end of 1996, five cases of PVP violation came to court: four in relation to fruit, one to wheat (interview Salvo, 1996).

The transnational crop development conglomerates are also active in the seed markets of Chile and Colombia. They supply all seed of hybrid varieties of maize, sorghum, sunflower, as well as seed of several other crops. The main American and European hybrid maize seed enterprises also have production sites located in Chile, where seed for export is produced during the wintertime of the Northern hemisphere. Their advantage from PVP is marginal as the hybrids have a built-in protection against unauthorized propagation, and as production takes place under strict corporate control. Rather than PVP, the conglomerates may in due course use the opportunities under the patent laws of Chile and Colombia to protect genetic information incorporated in their hybrids against unauthorized exploitation.

6.3.3 Opposition to plant-related IPR

Because PVP reduces the opportunities to trade on-farm saved seed informally, and patent coverage for plant varieties even makes all on-farm propagation illegal, farmers are the most obvious potential opponents of plant-related IPR. Nevertheless, IPR protection has not yet been an issue for the major farm organizations in Chile (interviews Orella, 1996; Molina, 1996) nor in Colombia (interviews Arevalo and Luján, 1996; Gómez, 1996; Sanabria and Dabalos, 1996; Quintama and Pordnía, 1996). Farm organizations have not been involved in the introduction of PVP or the patent law changes, nor have they been represented at the international forums where plant-related IPR has been at issue, such as the CBD. One of the reasons is that several of the organizations, such as FEDEARROZ and FEDECAFÉ in Colombia, and SNA in Chile, also have plant breeding programmes. In respect of PVP, they therefore have mixed interests.

Another reason for the lack of interest in IPR by farm organizations is that the effects of plant-related IPR protection on farmers vary greatly. Two groups of farm systems will probably hardly notice a negative impact: the capital-intensive farms and the peasantry. The capital-intensive farms predominantly use certified seed, purchased from breeders or their official dealers. They are used to paying official prices for the most technologically advanced varieties, and have the means to employ them in a profitable way. The Chilean and Colombian peasants, or *campesinos*, presumably also remain relatively immune from IPR protection of plant varieties, as they hardly use any modern commercial varieties. Even if they did propagate varieties protected under patent law, it would be almost impossible to detect this propagation, and presumably not worth the cost.

Most affected by plant-related IPR protection are the middle-sized farms, particularly those that are struggling for survival as commercial undertakings in the process of agricultural industrialization. In order to reduce production costs, these farms may produce seed on-farm and obtain a part of their seed requirements from informal channels. Informally traded seed is cheaper because payment of royalties and, more importantly, taxes can be evaded. It is precisely these informal seed distribution channels, however, that are the target of PVP. The formalization of these channels reduces the opportunities to obtain high quality seed at an advantageous price.

Nevertheless, the major farm organizations have not opposed legal protection of plants. The reason must presumably be sought in the fact that the concept of IPR coverage for plant material is entirely new to the farm circles in the Latin American countries. PVP has only recently been enforced so that farmers have not yet encountered its effects. The same is true for plant patenting. In the mid-1990s, the opportunity to patent plant material was new even to most of the plant breeders we interviewed.

The organizations that did oppose PVP and plant patenting in Chile and Colombia include nature and farm-oriented NGOs, and organizations of indigenous peoples. They are also the organizations that favour an alternative agricultural production strategy, which is non-industrial and farmer-oriented (*cf.* Anonymous, 1995b; Anonymous, 1996b; Castaño, 1996; Del Pilar Valencia, 1996; Flórez, 1996; Manzur, 1995a; 1995b; Montecinos, 1996; Pombo, 1996). The organizations reject the principle of assigning IPR to biological organisms. They maintain that the concept of IPR protection is alien to the local farm and indigenous communities, and contrary to their cultural and religious values. The organizations state that the introduction of IPR protection disrupts the traditional uses of genetic resources of plants. By assigning private rights over plant varieties, the farmers' and indigenous communities in their perception "lose control" over the use of their plant genetic resources. They do not benefit from the exploitation of their landraces and wild relatives, while IPR protection also restricts or eliminates the opportunity for farmers to save and exchange their seeds. There is also concern that even the use of traditional varieties will be prohibited if they carry genes that are covered by a patent.

The organizations also reject the international division of labour in crop development. The introduction of the Northern IPR legislation in Chile and Colombia facilitates the presence of the transnational crop development conglomerates in the national seed markets. The conglomerates bring along their uniform and homogenous plant varieties, which replace traditional varieties. The resulting loss of traditional diversity, both at the varietal and genetic level, is considered to be dangerous for national food security. The reliance on foreign, uniform varieties makes agricultural production more susceptible to pests and diseases, and increases the use of imported agro-chemicals. Moreover, IPR protection adds a 'capital drain' to the 'gene drain' as domestic breeding firms and producer organizations must pay royalties to use the foreign varieties. Licence conditions for such technology transfer may further limit the possibilities for using the germplasm.

Most organizations opt for an alternative IPR system that protects collective intellectual property rights of peasant and indigenous communities over their resources, their traditional knowledge and innovations. These communal rights are the basis for negotiating a share of the benefits derived from exploitation of community resources by third parties. Several organizations point out that, at least as far as indigenous communities are concerned, the compensation should be paid through, for example, the establishment of a better health care system. The payment of money may cause more harm than good. It may provoke conflicts among communities and disrupt the traditional way of life (interviews Aguilar, 1994; Martínez, 1996; Green, 1996).

Other organizations promote a more radical strategy. In their publications, Montecinos (1996) and Castaño (1996) maintain that "the alternative of IPR is no IPR." They argue that entering into the so-called bio-prospecting contracts implies an acceptance of the patenting of native genetic resources, with the negative consequences for the socio-economic and cultural life of rural population as outlined above.

None of the IPR-opposing organizations in Chile and Colombia treat IPR protection as an isolated issue. They consider PVP and patent coverage of plant varieties to be elements of a particular agricultural production strategy embedded in an agro-industrial division of labour that they reject and do not want to be part of. This conviction is most clearly expressed by Montecinos (1996:25) who favours a total disruption of economic internationalization: "The only way to fight those [Northern industrial - RP/JvW] monopolies is by fighting mega-markets. This will happen when we reclaim diversified production systems based on local resources and knowledge, when agriculture ceases to be an input-consuming machine, when farmers reclaim the right to use and develop their own technology, when we stop eating the same thing in Manilla, Pittsburg and Concepción, and our health no longer depends on Monsanto and Ciba Geigy. This will also be the only way in which indigenous peoples and farmers will become the primary beneficiaries of what their societies have created and continue to create".

Not all nature and farm-oriented NGOs and indigenous people's organizations in Chile and Colombia would agree with this total rejection of the international agro-industrial division of labour. However, they do favour an alternative agricultural production strategy directed at the autonomous improvement of subsistence conditions in the poorest rural areas. And rather than Western cultural elements, the local, traditional culture and religion is to play an important role in this.

This is indeed also the strategy being followed by peasant and indigenous communities. These two types of communities, which partly overlap, may have their own reasons for doing so, however. The peasants are less guided by ancient knowledge and religious rituals in their methods of agricultural production than indigenous people. They produce with little industrial inputs, use landraces or old varieties and on-farm saved seed, not because they intentionally strive for a non-industrial production strategy or because of a desire to conserve biological diversity, but because they simply do not have the means to purchase industrial inputs.

This is presumably the reason that the two peasants' organizations we interviewed, *La Voz del Campo* in Chile and the *Asociación Nacional de Usarios Campesinos* (ANUC) in Colombia, were hardly interested in the IPR dispute in their respective countries. For both organizations the term 'Farmers' Rights' had a connotation which had little to do with the compensatory mechanism that has been agreed upon in the FAO. Farmers' rights were rather interpreted as the "inalienable right" of farmers to have access to land, credit and support, to relieve hunger and to live in peace (interviews Gómez, 1996; Molina, 1996). There may be rural resentment against those companies that benefit from using landraces, but obviously the peasantry in the two Latin American countries have problems that are more pressing than compensation for this 'bio-piracy'.

The awareness of IPR was far greater among the indigenous communities of Chile and Colombia, partly because the widely publicized cases of the patenting in the USA of DNA of a number of indigenous people from Latin American and Asian countries, and the Human Genome Diversity Project. Tapping a person's blood, and assigning private rights for the exploitation of knowledge is entirely at odds with the culture of the indigenous peoples. The communities often attach spiritual value to blood, plants and other biological material. Moreover, the knowledge they have accumulated within their communities is closely linked to traditions. In Colombia, the communities in Antioquia may collaborate with external researchers, but only on their terms. Patenting inventions based on material collected in their territory is not allowed (interview Martínez, 1996). However, the intentions of indigenous communities should not be romanticized. They will also use modern plant varieties if these are affordable and have proven to be better than the traditional ones (interview Contreras, 1996). The IPR and access issues provoked fierce reactions among the indigenous communities in Latin America, because they come on top of a lot of other problems of a more structural character, such as territorial disputes and ongoing discrimination against indigenous peoples in Western society (interview Herrera, 1996).

6.3.4 Rival agricultural production strategies provoke IPR controversy

In the previous section we have shown that PVP in Chile and Colombia has both been embraced and rejected. Domestic breeding and producers' organizations, which are in favour of a further industrialization of agriculture welcome the introduction of PVP, albeit for divergent reasons. Producers of non-traditional export commodities require PVP to facilitate access to the foreign varieties primarily of cut flowers and fruit. Legal protection will not incite them to invest in breeding. They consider the technology gap in relation to the foreign breeders to be too wide and have no long-term confidence in the stability of the political situation in their country. Colombian producers of traditional export commodities, such as coffee and sugar cane, also favour PVP. They have their own breeding programmes, and PVP may help them to defend Colombia's technological edge compared to competing producers in the Central and Latin American region. Domestic private breeding firms or producer organizations with their own breeding programmes on crops directed at the national market, such as wheat in Chile and potato and rice in Colombia, also strongly support PVP. Legal protection against unauthorized propagation of their varieties enables them to defend their share of the seed market, which is shrinking due to economic liberalization, periods of drought and rural violence. Finally, the national agricultural research institutes wish to use PVP to compensate for the shrinking budgets. Royalties are considered to be a new, additional source of income.

The introduction of PVP advances the industrialization of crop development. PVP is adopted to encourage private capital investment in plant breeding in order to generate more productive and adapted varieties. PVP, however, is only one ingredient in the cocktail of industrial crop development. Seed laws and additional farm policies are designed to make farmers shift from their landraces and informally produced seed, to the modern and commercial varieties. In most developing countries the use of certified seed is either recommended by extension services, required as a condition for the use of credit facilities, or obliged by the processing industry.

The industrialization of crop development is also promoted by the transnational crop development conglomerates, which have been the dynamic force behind the worldwide strengthening of plant-related IPR. The transnational conglomerates and the breeding organizations in Chile and Colombia therefore have a mutual interest in IPR, albeit with a crucial difference. The conglomerates benefit primarily from patent protection, not from PVP. They hardly need PVP, because they can prevent unauthorized propagation by selling hybrid varieties, and recently also by applying so-called 'terminator' technology. The conglomerates would rather benefit from opportunities to patent their plant genetic information in Chile and Colombia. They are among the very few organizations in the world that have the financial and technological capacity to generate plant-related *inventions*. Thus, while producers of non-traditional export commodities and some staple crops in Chile and Colombia already rely on varieties bred by the OECD-based crop development industry, the conglomerates are likely to strengthen their position in Chile and Colombia in relation to domestic breeders, either by selling hybrid varieties in more crops, by applying 'terminator' technology, or by seeking patent coverage for plant varieties.

Opposition to both PVP and patent protection of plants comes particularly from the nature and farm-oriented NGOs and indigenous people's organizations. These organizations act as a last resort for all those farming and indigenous communities that cannot participate in the process of agro-industrialization. They disapprove of plant-related IPR because it is meant to encourage private investment in crop development, and thus furthers the agro-industrialization process, which is considered to be the main contributor to the marginalization of the peasantry. In a similar vein, the organizations object to the collection of diverse biological organisms from the local communities' territory. These organisms are eventually exploited by enterprises that belong to the core of the agro-industrial system, which has little to contribute to non-industrial farming. The benefits derived from the locally collected natural resources therefore cannot be enjoyed by the peasants. The lack of land, capital, infrastructure and markets prevent them from truly benefiting from the world's advanced plant varieties. Just receiving a share of the possible financial revenues can by no means compensate for the farm communities' continuing isolation from the agro-industrial system.

In their rejection of the strategies directed at industrialization of agriculture in general and the IPR system in particular, the Southern NGOs are supported by nature and farm-oriented NGOs in the North. The Northern NGOs oppose plant-related IPR for ethical and environmental concerns, or because the seed-saving opportunities for Northern farmers may be restricted by IPR. Both Southern and Northern NGOs form a transnational alliance against the granting of intellectual property protection of biological material, in the same way that they form an alliance over biodiversity issues. Nevertheless, a basic difference remains between these NGOs. Since virtually all farms in the advanced industrialized countries are incorporated in the agro-industrial system, there is almost no social base for advancing the non-industrialized, farmer-oriented strategy in these countries themselves.

Summary

Basically two positions on the introduction of plant-related IPR can be distinguished in Chile and Colombia. Plant breeding organizations and producers of non-traditional export commodities advocate IPR. The breeders hope to reduce the unauthorized propagation of their varieties in Colombia and Chile, as well as in neighbouring countries; the producers focus on an improvement of the conditions for importing foreign plant varieties. Meanwhile, opposition to IPR protection is voiced by nature and farm-oriented NGOs and indigenous people's organizations. In this section it has been argued that this opposition is in fact not merely related to the direct impact of IPR protection on the peasantry, but to the entire process of agro-industrialization. Plant-related IPR protection is one element of a production strategy that is directed towards the industrialization of agriculture, a development that has little to offer resource-poor farm communities, which cannot participate.

Conclusions

This chapter sought to examine the roots of the international PGR conflict, not from a North-South perspective, but from the perspective of conflicting agricultural production strategies. Rather than depicting the conflict simply as an unequal exchange between the 'gene-rich and technology-poor South' and the 'gene-poor and technology-rich North', we have argued that the conflict is far more transnational in origin.

We have distinguished between three alternative agricultural production strategies advocated in OECD as well as in developing countries. In most countries, rival opinions on the most preferable organization of agricultural production and progress also find expression in conflicts on the exploitation and conservation of plant genetic information. Hence, the 'PGR conflict' is not a new, isolated phenomenon in international relations, but a 'spark' resulting from the frictions between rival agricultural production strategies. As these strategies have a transnational dimension, the conflicts between them also tend to cut across countries and regions. This explains why the PGR conflict is 'transnational' in scope, while the 'front lines' overlap those that divide industrialized and non-industrialized farm systems.

The opposition to the gene-drain and the patenting of plants and other biological material comes particularly from nature and farm-oriented NGOs and indigenous people's organizations. These organizations act as a last resort for all those farming and indigenous communities that cannot participate in the process of agro-industrialization. The communities seem to have little to gain from the emerging international division of labour in crop development. They lack the land, the appropriate soil, the capital, the markets, and the infrastructure. Despite national rural development programmes, these resource-poor farms cannot be converted into 'viable' commercial family farms. Under neo-liberal policies, the support of 'non-viable' farms is considered to be an inefficient allocation of government means. In this way, the poorest category of farm communities is increasingly dependent on national and foreign aid programmes. They are further marginalized from the market economy and left to their own, non-industrial production and survival strategy.

Traditionally, the interest in this non-industrial agricultural production strategy on the part of private and public research organizations has been low. The organizations that collect landraces and wild relatives, and protect their varieties with IPRs, form part of the transnational, agro-industrial system, which has little to contribute to the non-industrial production strategy. The plants are collected with the objective of improving market-oriented agro-food production and will hardly contribute to non-industrial agriculture. Benefits derived from the collected natural resources cannot be enjoyed by the peasantry, because they lack the means to industrialize their agricultural production. A share of the possible financial revenues cannot compensate for the fact that the farm communities are isolated from the agro-industrial system. Plant-related IPR is meant to encourage private investment in crop development. This implies a furthering of the agro-industrialization process, which is considered a strong contributor to the marginalization of the peasantry.

The intensity of the controversy on conservation and plant-related IPR protection is not only determined by their spatial, transnational aspects, but also by a historical dimension. As discussed in chapters 2 to 5, the development of conservation and IPR regulation has been crucial in paving the way for an unprecedented agro-industrial development in the past 100 years. The chapters show that the adoption of this regulation could only take place on condition that the majority of far-mers were included in an industrialized production system. This important precondition is not being met during the implementation of conservation and IPR regulation in developing countries. At the end of the 20th century, both types of regulation are being dropped like alien spaceships into regions where farmers do not have the means to adopt industrialized forms of production. They do not view plants as sources of genetic information or 'genetic resources'; neither are they accustomed to Western, individual property rights. Solutions to the PGR conflict are therefore most likely to be found through a thorough understanding of the relationship between industrialized and non-industrialized agricultural production systems.

Notes

1 Monsanto's CEO Verfaillie (1997) remarks: "Why is Monsanto interested in sustainability? The simple answer is that we believe current models of economic growth are a dead-end street. We know the human race can't go consuming resources the way we have for generations, due to absolute constraints on the earth's biological and physical systems ... We must acknowledge that there are limits to what the earth can provide." DuPont's Internet homepage on "safety, health and the environment" contains a similar message. "Recognizing that the Earth's resources are limited and that conservation is necessary for human survival, many with diverse views are coming together to work on the issue of common interest ... Only those companies and industries that provide value to society in a way that is protective of the world's resources will be allowed to operate into the 21st century." (www.dupont.com/corp/gbl- company/she/views.htm#sustain)

2 The International Agri-Food Network promotes cooperation between major international agro-food organizations, including the international plant breeders' association (ASSINSEL), the International Seed Federation (FIS), the Global Crop Protection Federation (GCPF), the International Fertilizer Industry Association (IFA), and the International Federation of Agricultural Producers (IFAP) (IAN, 1998).

3 In Chile, around 15% of the population live in rural areas. In Colombia this figure is 25%.

4 Romanticism in this respect refers to a conception of nature, rooted in Western culture protest, that goes back to Jean-Jacques Rousseau, who is known for his ideal of the pure, untouched natural environment as a condition for an uncorrupted society (*cf.* Grove, 1995:224).

5 See especially Art. 63 and 330 of the Colombian Constitution (Republica de Colombia, 1993).

6 In Chile, plant variety protection (PVP) legislation was enacted in 1994 (Ministerio de Agricultura, 1994), and in Colombia in 1995 (ICA, 1995). The Chilean PVP legislation replaced a previous, hardly enforced, PVP system, which had been adopted under the military regime in 1977. In Colombia, PVP was introduced through an Andean Pact decision to establish a common regime on PVP, in October 1993 (Comisión del Acuerdo de Cartagena, 1993a). In Chile a new patent law was enacted in 1991 (Ministerio de Economía, Fomento y Reconstrucción, 1991); Colombia's patent law was modified on the establishment of a common regime on industrial property protection among the Andes Pact

countries in 1993 (Comisión del Acuerdo de Cartagena, 1993b). Plant varieties are explicitly excluded from patent protection in Chile, but explicitly included in the Colombian law.

Conclusion

The collection and conservation of plants and seeds in developing countries and the legal protection of new plant varieties have provoked considerable controversy over the past decades. International agreement has not yet been achieved, as evidenced by the difficulties experienced in the conclusion of two major international agreements negotiated in the 1990s: the Convention on Biological Diversity (CBD) and the agreement on Trade-Related Aspects of Intellectual Property Rights (TRIPS). The crux of the disagreement is the question of who is to benefit from the exploitation of plant genetic information. For some, the transnational enterprises that exploit and protect plant genetic information collected in the developing countries are guilty of 'bio-piracy'. Others shrug off this criticism, pointing out that plants are a renewable resource, and that a careful collection of plants does not harm farming opportunities in developing countries. It is also often contended that domestic enterprises in developing countries are located in the midst of diverse genetic resources without exploiting them. If these enterprises do not themselves take advantage of the biodiversity, it is argued, why should they prevent foreign enterprises from doing so?

In this book we argue that the root of the 'PGR conflict' must be sought in the industrialization of agriculture. Agro-industrialization can be considered as the fundamental dynamics in agriculture and comprises the gradual transformation of farming into an industrial production process. Step by step, discrete farming activities are being replaced by industrially manufactured goods; horses are replaced by tractors, manure by fertilizer, landraces by high-yielding plant varieties, and on-farm saved seed by commercial seed. In some cases, industrially produced foodstuffs entirely replace farm produce.

Since the pace of agro-industrialization differs per country and region, the competitiveness of national agricultural sectors varies greatly and induces an international division of labour in agro-food production. The Third Agro-Food Order, which emerged in the 1980s, is based on an international division of labour in agriculture in which the OECD countries prevail. They are the main exporters of cereals, fruits, vegetables and flowers, while the developing countries, as a whole, are the main food importers. The OECD countries are also the world's major exporters of advanced, commercial plant varieties. This situation is due to an international division of labour in crop development itself, with industrial conglomerates, based in OECD countries, constituting the nucleus.

We have shown that a country's position in the international division of labour in agriculture is closely linked to its position in crop development. In order to produce competitively, individual countries are required to participate in the international division of labour in crop development, and to follow the internationally agreed rules regarding the conservation and exploitation of plants. For this reason, they have to design conservation policies that connect their national seed collections to the international conservation system, and they have to restrict unauthorized exploitation of plant varieties by accepting the various mechanisms that protect imported, foreign plant varieties.

The prime beneficiaries of the international regulatory framework seem to be the transnational crop development conglomerates. They are among the very few actors who have the financial and technological capacity to exploit the global genepool and to protect modified plant genetic information. Commercial producers in developing countries also have an advantage, but it is indirect. Acceptance of the international rules offers them opportunities to import better plant varieties. Plant breeding organizations in developing countries may benefit from the regulation, as they can protect their own plant varieties against unauthorized propagation. At the same time, however, their position vis-à-vis the conglomerates tends to weaken, because the opportunities to patent plant genetic information reduces their opportunities to exploit the pool of varieties freely for further breeding.

The major opponents of the international crop development regulation are the nature and farmer-oriented non-governmental organizations (NGOs) and indigenous peoples' organizations. They act as last resort for all those small-scale farmers who cannot participate in the process of agro-industrialization, because they lack the land, the credit, the infrastructure, and the markets. In principle, the peasantry could play a significant role in the international division of labour in crop development as suppliers of plant genetic information. Much of the world's biodiversity is found in, or around the peasants' fields, and in the territories of indigenous peoples. However, they have little to win from international crop development policies that have the explicit objective to promote industrialized forms of agriculture. The number of so-called 'non-viable' farms, unable to keep pace with the agro-industrialization process and left to fend for themselves, is increasing. Unlike in OECD countries, there is little prospect of alternative employment in the industrial or service sectors, while adequate social security schemes are generally absent. As a result, the industrialization of agriculture and the associated international crop development regulations are considered as a threat to resource-poor farmers. The PGR conflict is therefore likely to intensify, irrespective of the regulations that are being developed to compensate developing countries or their rural population for the supply of plant genetic information. The sharp disapproval of the recently disclosed 'terminator technology' may serve as an illustration. The technology has been denounced as a "genetic bomb", a "Trojan Horse", and as "biological warfare" (RAFI, 1998c).

Every considered option to ease the PGR conflict must take into account the fundamental contradiction in policies that foster agro-industrialization. On the one hand, a further industrial transformation of farming seems to be unavoidable in view of an expanding, basically urban world population. On the other hand, the same process increasingly excludes parts of the rural population from participation in agriculture, forcing them to migrate to the cities. The management of this contradiction requires that the agro-industrialization process be led not by market prices alone, but also by active steering on the basis of national political agendas that mitigate or prevent undesirable consequences. Such political intervention is not just in the interest of the peasantry, but also of governments in developing countries, and of the transnational crop development industry.

Because of their innovative capacity and influence in the design of international conservation and IPR regulation, the industrial conglomerates have become central actors in crop development and agriculture worldwide. Their influence is so substantial that the conglomerates, such as Monsanto, DuPont-Pioneer, ELM-Pulsar, Novartis, Rhone Poulenc-Limagrain and Zeneca-Cosun, can no longer ignore the social opposition their strategies arouse. On the one hand, this is a question of ethics. Their key position in farming and food production bestows on the industrial conglomerates with a public responsibility to farmers and consumers. On the other hand, however, it is a matter of sheer corporate interest. Large global sales are the foundation of the conglomerates' influential position, but also make them vulnerable to civil resistance.

For example, local farming and indigenous organizations may hinder industrial exploitation of locally available plants. By challenging the authority of the national PGR curator of their country, they could obstruct the national biodiversity access regulation, designed by the central government, as agreed upon under the Convention on Biological Diversity. Encouraged by Greenpeace's success in opposing the intended dumping of Brent Spar by Shell, farmers and consumers' organizations may also use their purchasing power as a trump card in their opposition. Industrial crop development conglomerates also run the risk of violent opposition. Their personnel and products may increasingly be received with hostility by parts of the rural population in developing countries. In this respect we refer to the ransacking of the local Cargill office by farmers in India, some years ago. In order to diminish the chances of being the victim of such actions, Pioneer Hi-Bred Colombia has the policy of hiding its office. When we visited the plant, there were no signboards carrying the company's name, and the location of the company was hardly known among the local population. The marginalization of an increasing part of the rural population in developing countries may have devastating effects. This process undoubtedly supports political agitation and instability, and encourages migration to the slums of national cities or to OECD countries. All these developments are taking place now, and potentially have not only regional but also worldwide consequences.

Now the question arises whether or not it is possible for governments and industrial conglomerates to adjust their strategies in order to reduce the high social costs associated with the process of agro-industrialization. We think it is possible, provided that a number of requirements are met.

The first requirement is that CEOs and governmental leaders reconsider their one-dimensional, global view on agricultural production and crop development. The metaphor of the 'global village' that is often employed to illustrate that adequate decisions can be taken at a global level about agro-food production and distribution, disguises the many national and regional differences. The uneven advances in the process of agro-industrialization, combined with the vast socio-economic and cultural diversity in world agriculture, make it impossible for agro-food production to be served from a central position, with a single uniform regulatory recipe. International crop development regulation may fit all those agricultural sectors - in OECD as well as in developing countries - that are incorporated in, and benefit from industrialized production systems. However, in most developing countries, a large part of the rural population pursues a basically non-industrial and non-commercial agro-food production strategy. It does not make sense that international crop development regulation is applied to, or interferes with, this type of production.

In the second place, it is essential that governments and CEOs recognize this non-industrial survival strategy, and consider production improvement by non-industrial means as a goal in itself. Instead of maintaining a dismissive attitude towards "backward" forms of production that should already have been replaced, the use of traditional varieties, on-farm saved seed, organic fertilizer and biological insecticides must be considered as a viable and inevitable production alternative that deserves professional support. Science should be employed to enhance the exploitation of resources readily available in the region, on condition that the knowledge, habits, preferences and ethics of the local communities are taken as a starting point.

The necessity to support non-industrial agriculture has already been acknowledged in recent policy papers of the international agricultural establishment (*cf.* CGIAR, 1998; Conway, 1997; Conway *et al.*, 1995; Srivastava *et al.*, 1996). In these publications it is recognized that, apart from an intensification of production in high-potential areas, more research is needed in lower-potential areas. The aims of this research should be the improvement of farming systems rather than specific commodities, a reduction in the use of external inputs (pesticides and fertilizers), and greater involvement of farmers and local communities in the research design.

Even though this policy shift is to be welcomed, the critical reaction it has provoked on the part of some farm-oriented NGOs, notably the Rural Advancement Foundation International (RAFI, 1998d), seems justified. The main international agricultural research centres and the CGIAR were the central engine of the Green Revolution, which facilitated the agro-industrialization process in developing countries. It is therefore not obvious that these organizations should take the lead in the design and implementation of research in support of non-industrial agriculture. Most plant breeders and molecular biologists have been educated in the idea that agricultural improvement is determined by technological opportunities, rather than by farming and ecological capacity. In order to be able to facilitate low external input agriculture, they must first radically change their attitude towards non-industrial farming and farmers. This can only be achieved if a new, modern appeal is attributed to close peasant/scientist relationships and to many farming elements that are usually regarded as laborious, obsolete, traditional, and primitive. Those non-governmental organizations that are experienced in research on non-industrial agriculture in developing countries should be given a prominent role in this cultural transformation of scientists.

The third requirement for governments and industrial conglomerates to reduce the high social costs associated with the process of agro-industrialization and to ease the PGR conflict is an adjustment of the international crop development regulation itself. Below, we map out a number of specific options for such changes in regulation with respect to conservation, biodiversity access and plant-related IPR protection.

Conservation

It is necessary to remove the double agenda that currently governs global conservation efforts, financed by OECD countries. In whose interest is the global conservation system in the first place? Obviously the crop development conglomerates are among the very few that have the resources and know-how to 'tap' the global conservation system for commercially interesting genetic information. The frequently heard argument that some peasant communities may benefit just as much is not entirely false, but should not disguise the fact that the system was not designed for them. The global conservation system is to support agro-industrialization and thus entails little benefit for non-industrial agriculture. To repair the broken relationship between local peasant interests and global conservation goals, it is necessary to design an additional, alternative conservation strategy that re-establishes the closed circle of conservation, local breeding, and agricultural production. Linking conservation to local production should be recognized as a goal in itself and not necessarily be related to the global conservation system.

Access regulation

Governments that are in the process of implementing the CBD tend to evade questions of access to and compensation for genetic information. An important reason is that a debate on these questions fans the flames of other more controversial issues, such as ethnic and territorial rights. Many governments, therefore, prefer a very diplomatic and 'soft' interpretation of the Convention. The fact that marginalized farmers' groups and indigenous peoples' organizations use the CBD as a political 'spring board' is, nevertheless, an important sign on the wall. It indicates that the heated debate on fairly technical access and compensation issues cannot be understood in the context of global environmental protection, but only in the socio-economic setting of marginalized farmers and indigenous groups. In Chile and Colombia, governmental attention for marginalized groups in the implementation of the CBD is confined to their role as 'custodians' of biodiversity, and not as representatives of an alternative, viable, non-industrialized agricultural production system. As long as non-industrialized farmers are not recognized as viable agricultural producers, they will continue to 'stir up' the CBD implementation process.

Governments may benefit from a dual approach in the design and implementation of access regulation. The starting point for a dual approach is the recognition of the divergent interests of industrialized and non-industrialized farmers in access policies. Farmers who participate in industrialized, commercial agro-food production systems often rely on foreign plant varieties or on domestic varieties that incorporate foreign genetic information. They therefore have an interest in a free access regime, thus allowing foreigners to use national biodiversity. A restrictive access policy of their own government might provoke reciprocal actions from governments of countries where they obtain their varieties. The free access approach for industrialized producers implies that, for example, Pioneer Hi-Bred is allowed use CIAT's or CORPOICA's maize collections as long as these two institutes are convinced that they will receive sufficient genetic information in return, be it in the form of DNA sequences, finished varieties, or landraces from Pioneer's own collections.

The dual approach, however, implies that farmers involved in non-industrial production, who have no *direct* interest in global conservation, are free to interpret access regulation in accordance with their interests. A strict access regulation may help individual farming or indigenous peoples' communities to generate funds from foreign institutes or enterprises, as in the Merck-InBio 'bioprospecting' agreement. Monetary compensation, however, is not sufficient to counter the process of marginalization to which most non-industrialized farmers are subjected. Bioprospecting agreements should therefore be explicitly related to both the recognition and the refinement of non-industrial agriculture. In practice this implies that bioprospecting contracts should be linked to the pressing needs of the rural population in the region where the plants are being collected. These needs are access to land, credit, markets, infrastructure, health, education, empowerment, and non-industrial crop development.

Plant-related IPR protection

The dual approach to access regulation should be extended to plant-related IPR protection. It seems fair that plant variety protection (PVP) should be implemented and enforced in industrialized agro-food production sectors in developing countries, such as that of cut flowers in Colombia and of fruit in Chile. It is undesirable, however, that PVP should hinder propagation of varieties by the peasantry and those categories of medium-sized farmers who are economically most vulnerable. To this end, national governments deserve support for their attempts to retain opportunities to exclude certain crops from PVP, or to exempt on-farm saving and the exchange of seed among resource-poor farmers. In this respect, the PVP and biotechnology patent legislation adopted in the European Union may serve as an international precedent. It also seems justifiable that governments retain the right to control the growing of hybrid varieties or varieties that contain the "terminator" gene. While such varieties are acceptable for industrial farming systems, they are inappropriate for (and may negatively affect) the majority of the rural population in developing countries.

So far, plant-related IPR has primarily attracted the attention in developing countries of farmer and environment-oriented NGOs, in view of its negative impact on farmers. Plant breeding organizations in developing countries have, to a certain extent, considered the impact of PVP on their future breeding activities, but not the effects of patent coverage - an issue they should certainly address. Patent protection of whole plants, plant varieties and their progeny entails the real danger that in the near future a large part of the genepool of commercial varieties will no longer be freely available to plant breeders. For this reason, it seems justified to consider limiting the scope of the patent claims for plant-related inventions to techniques or specified genetic information.

References

Abercrombie, M., Hickman, C.J., Johnson, M.L. (1980) *Dictionary of Biology* (seventh edition). Penguin Books, London, United Kingdom.

Addens, N. (1952) *Zaaizaad en Pootgoed in de Nederlandse Landbouw* [Seeds and Seed Potatoes in Dutch Agriculture]. Veenman & Zonen, Wageningen, the Netherlands. (summary in English)

Advanta (1997) Key figures & addresses 1997. Advanta, Capelle, the Netherlands.

Aglietta, M. (1979) *A Theory of Capitalist Regulation. The US Experience.* New Left books, London, United Kingdom.

Aglietta, M. (1982) World capitalism in the eighties. *New Left Review* 136, 5-41.

AIPPI (1933) *Congrès de Londres.* Annuaire de L'Association Internationale pour la Protection de la Propriété Industrielle, 36(4), 2e série 1932.

AIPPI (1954) *Congrès de Bruxelles.* Annuaire de L'Association Internationale pour la Protection de la Propriété Industrielle, 57(5), 3e série 1954.

Alston, J.M., Pardey, P.G., Smith, V.H. (1998) Financing agricultural R&D in rich countries: what's happening and why. *The Australian Journal of Agricultural and Resource Economics* 42(1), 51-82.

Andrews, D., Mitchell, M., Weber, A. (1979) The Development of Agriculture in Germany and the UK. Comparative Time Series 1887-1975, Issue No.3. Centre for European Agricultural Studies, Wye College, Ashford, United Kingdom.

Anonymous (1991) Actions taken by FAO conference reflect new era of harmony for international plant genetic resources. *Diversity* 7(4), 4-7.

Anonymous (1992) Political and economic strife around the world causes gene banks to face loss of irreplaceable collections. *Diversity* 8(3), 13.

Anonymous (1995a) "Yo de presidente daría una orden de comando: hay que fomentar la industria de espárragos ya". *Estrategia* 15 February, 22-25.

Anonymous (1995b) Recursos genéticos. ¿Para quién? *Semillas en la Economía Campesina* 4, June 1995, 3-8.

Anonymous (1996a) Dinamizar el mercado. *Estrategia* 15 August, 16.

Anonymous (1996b) Accesso a los recursos genéticos, protección de derechos colectivos y securidad alimentaria. *Semillas en la Economía Campesina* 7, June 1996, 4-8.

Anonymous (1997) An all-you-can-eat genome project. *ScienceScope* 277(5328).
Anonymous (1998) The invisible giant casts a large shadow over CBD: the non-party status of the United States. *Diversity* 14(1&2), 29.
Arts, B. (1998) *The Political Influence of Global NGOs: Case Studies on the Climate and Biodiversity Conventions.* International Books, Utrecht, the Netherlands.
ASOCOLFLORES (1994) ASOCOLFLORES 38, March 1994. Asociación Colombiana de Exportadores de Flores, Bogotá, Colombia.
ASSINSEL (1947) *Official Report of the 2nd Congress of 'Assinsel', Paris, 27th, 29th and 29 of June 1947.* Association Internationale des Sélectionneurs pour la Protection des Obtentions Végétales, Paris, France.
ASSINSEL (1948) *Procès-Verbal du 3me Congrès de l'Assinsel, Scheveningen, Hollande, 21 au 24 Juin 1948.* Association Internationale des Sélectionneurs Professionnels pour la Protection des Obtentions Végétales, Paris, France.
ASSINSEL (1949a) *Procès-Verbal du Conseil de l'Assinsel qui s'est tenu à Amsterdam, les 28 et 29 Novembre 1949.* Association Internationale des Sélectionneurs Professionnels pour la Protection des Obtentions Végétales, Paris, France.
ASSINSEL (1949b) *Minutes of the Meetings. October 28th and 29th, 1949, January 7th 1950.* International Association of Professional Selectioners for the Protection of Vegetal Obtainments. Paris, France.
ASSINSEL (1950) *Procès-Verbal du 5e Congrès du 23 au 25 Juin à Paris.* Association Internationale des Sélectionneurs Professionnels pour la Protection des Obtentions Végétales, Paris, France.
ASSINSEL (1954) *Report of the 6th Congress, From June 13th to June 16th in Brussels.* Association Internationale des Sélectionneurs Professionnels pour la Protection des Obtentions Végétales, Paris, France.
ASSINSEL (1981) Feeding the 5 billion. Association Internationale des Sélectionneurs pour la Protection des Obtentions Végétales, Nyon, Switzerland.
ASSINSEL (1997) Feeding the 8 billion and preserving the planet. Association Internationale des Sélectionneurs pour la Protection des Obtentions Végétales, Nyon, Switzerland.
ASTA (1998) Resolution passed by the ASTA Board of Directors presented by the ASTA *Ad Hoc* Committee on Plant Genetic Resources. Internet homepage of the American Seed Trade Association, Washington D.C., United States. (http://www.amseed.com/documents/) (consulted December 1998).
Barghouti, S., Cromwell, E., Pritchard, A.J. (eds.) (1993) Agricultural technologies for market-led development opportunities in the 1990s. World Bank Technical Paper No. 204, World Bank, Washington, D.C., United States.
Baum, W.C. (1986) *Partners Against Hunger.* Consultative Group on International Research, World Bank, Washington, D.C., United States.
Bent, S.A., Schwaab, R.L., Gonlin, D.G., Jeffery, D.D. (1987) *Intellectual Property Rights in Biotechnology Worldwide.* Stockton Press, New York, United States.
Berlan, J.P. (1983) Breeders' rights and the patenting of life forms. Paper presented at the EASST Conference: 'Biology: Long Term Development'. Science Museum, London, United Kingdom.
Berlan, J.P. (1997) Political economy of agricultural genetics. In: Singh, R., Krimbas, K., Paul D., Beatty, P. (eds.) *Thinking About Evolution: Historical, Philosophical and Political Perspectives.* Cambridge University Press, Cambridge, United Kingdom.
Bijman, W.J. (1997) Herstructurering van de zaadindustrie: wereldwijde strijd om genen, tech-

nologie en toegang tot de keten. *Spil* 147/148, 33-38.

Bijman, W.J. (1999) How biotechnology is changing the structure of the seed industry. *International Journal of Biotechnology* 1(2) (forthcoming).

BIO (1998) Internet homepage of the Biotechnology Industry Organization. Washington, D.C., United States. (http://www.bio.org) (consulted January 1999).

Board for the Plant Breeder's Right (1946) Plant Breeder's Decree 1941. The Legal Status of the Plant Breeder and the Trade in Propagating Material in the Netherlands. Algemeene Landsdrukkerij, Den Haag, the Netherlands.

Bonefield, W. (1988) Reformulation of state theory. *Capital and Class* 33, 96-127.

Brockway, L. (1979) *Science and Colonial Expansion: The Role of the British Botanic Gardens.* Academic Press, New York, United States.

Brockway, L. (1988) Plant science and colonial expansion: the botanical chess game. In: Kloppenburg Jr. (ed.) (1988b), pp. 49-66.

Brown, L. (1985) *State of the World.* World Resources Institute, Norton, New York, United States.

Brown, W.L. (1975) Maize germplasm banks in the western hemisphere. In: Frankel, O.H. and Hawkes, J.G. (eds.) *Crop Genetic Resources of Today and Tomorrow.* IBS series Vol. 2, Cambridge University Press, Cambridge, United Kingdom, pp. 467-472.

Brown, W.L. (1984) Conservation of gene resources in the United States. In: Yeatman *et al.* (1984), pp. 32-41.

Bugos, G.E. and Kevles, D.J. (1991) Plants as intellectual property: American practice, law, and policy in world context. Humanities working paper 144, California Institute of Technology, California, United States.

Bureau International de L'Union (1902) *Actes de la Conférence Internationale pour la Protection de la Propriété Industrielle. Réunie à Paris du 4 au 20 novembre 1880.* 2nd edn. Union Internationale pour la Protection de la Propriété Industrielle, Bern, Switzerland.

Bureau International de L'Union (1911) *Actes de la Conférence. Réunie à Washington du 15 Mai au 2 Juin 1911.* 2nd edn. Union Internationale pour la Protection de la Propriété Industrielle, Bern, Switzerland.

Bureau International de L'Union (1926) *Actes de la Conférence. Réunie à la Haya du 8 Octobre au 6 Novembre 1926.* Union Internationale pour la Protection de la Propriété Industrielle, Bern, Switzerland.

Bureau International de L'Union (1934) *Actes de la Conférence. Réunie à Londres du 1er Mai au 2 Juin 1934.* Union Internationale pour la Protection de la Propriété Industrielle, Bern, Switzerland.

Busch, L., Lacy, W.B., Burckhardt, J., Hemken, D., Moraga-Rojel, J., Koponen, T., De Souza Silva, J. (1995) *Making Nature, Shaping Culture: Plant Biodiversity in Global Context.* University of Nebraska, Lincoln, United States, and London, United Kingdom.

Busch, L. and Sachs, C. (1981) The agricultural sciences and the modern world system. In: Busch, L. (ed.) *Science and Agricultural Development.* Allenheld & Osmun, New Jersey, United States, pp. 131-156.

Butler, L.J. (1996) Plant breeders' rights in the U.S.: update of a 1983 study. In: Van Wijk and Jaffé (1996), pp. 17-33.

Butler, L.J. and Marion, B.W. (1985) The impacts of patent protection on the U.S. seed industry and public plant breeding. Food Systems Research Group Monograph 16, University of Wisconsin-Madison, Madison, United States.

Buttel, F.H. (1984) Biotechnology and agricultural research policy: Emergent issues. Cornell

Rural Sociology Series 140, Department of Rural Sociology, Cornell University, United States
Buttel, F.H., Kenney, M., Kloppenburg Jr., J, Tadlock Cowan, J. (1985) Biotechnology in the world agricultural system: a new technological order for the new technology? Cornell Rural Sociology Series 144, Department of Rural Sociology, Cornell University, United States.
Cailliez, B. (1997) The twenty leading companies in 1996. *Cultivar* May, 42-43.
Carew, J. (1964) Germ plasm control as it would affect variety improvement and release of vegetables and flowers. In: CSSA (1964), pp. 35-38.
Castaño, G. (1996) Aportes al panel: tendencias internacionales de la protección de innovaciones biotecnológicas y recursos genéticos. Paper presented at the Seminario Gestion de la Propiedad Intelectual en Biotecnología, May 6-11 1996, CIAT, Cali, Colombia. Instituto Mayor Campesino (IMCA), Buga, Colombia.
CCI (1994) *Análisis Internacional del Sector Hortifrutícola Para Colombia.* Corporación Colombia Internacional, Bogotá, Colombia.
CEBECO (1963) 25 Jaar Veredelingsbedrijf CEBECO. Emmeloord, the Netherlands.
Cecil, R. (1979) German Agriculture 1870-1970. The Development of Agriculture in Germany and the UK. Part one. Wye College, Ashford, United Kingdom.
CGIAR (1991) Plant genetic resources: vital for global development. CGIAR fact sheet 3. Consultative Group on International Agricultural Research, Washington, D.C., United States
CGIAR (1997) CGIAR 1996 Financial Report. Consultative Group on International Agricultural Research, Washington, D.C., United States.
CGIAR (1998) The international research partnership for food security and sustainable agriculture. Third system review of the CGIAR. Final report, 8 October 1998. CGIAR System Review Secretariat, Consultative Group on International Agricultural Research, Washington, D.C., United States.
Cochrane, W.W. (1979) *The Development of American Agriculture: Historical Analysis.* University of Minnesota Press, Minneapolis, United States.
Cochrane, W.W. and Ryan, M.E. (1976) *American Farm Policy, 1948-1973.* University of Minnesota Press, Minneapolis, United States.
Comisión del Acuerdo de Cartagena (1993a) Decisión 345 del 29 octubre de 1993: Régimen Común de Protección a los Derechos de los Obtentores de Variedades Végétales.
Comisión del Acuerdo de Cartagena (1993b) Decision 344: Régimen Común sobre Propriedad Industrial.
Comisión del Acuerdo de Cartagena (1996) Sumario del Decisión 391 del Acuerdo de Cartagena: Régimen Común sobre acceso a los recursos genéticos. *Gaceta Oficial* 12(213).
Commission of the EEC (1988) The agricultural situation in the Community. European Economic Community: 1988 report. Brussels, Belgium.
Commission of the EC (1998) Directive 98/44/EC of the European Parliament and of the Council of 6 July 1998 on the legal protection of biotechnological inventions. *Official Journal of the European Communities* L 213, 30 July 1998. Brussels, Belgium.
Conway, G. (1997) *The Doubly Green Revolution. Food For All in the Twenty-first Century.* Penguin Books, London, United Kingdom.
Conway, G., Lele, U., Peacock, J., Piñeiro, M. (1995) A vision for CGIAR: sustainable agriculture for a food secure world. A vision for international agricultural research. A statement by an external panel appointed by the Oversight Committee of the CGIAR. In: CGIAR, Renewal of the CGIAR: sustainable agriculture for food security in developing countries. Ministerial-Level Meeting, Lucerne, February 9-10, 1995. Background Docu-

ments on Major Issues. Consultative Group on International Agricultural Research, Washington, D.C., United States.
Cosun (1998) Cosun Internet homepage. (http://www.cosun.com) (consulted January 1999).
Cotter, J. (1994), The Rockefeller Foundation's Mexican agricultural project: a cross-cultural encounter, 1943-1949. In: Cueto (ed.) (1994), pp. 97-125.
Cox, T.S., Murphy, J.P., Goodman, M.M. (1988) The contribution of exotic germplasm to American agriculture. In: Kloppenburg Jr. (ed.) (1988b), pp. 114-144.
CPRO (1998) Suikerbiet wordt Fructanbeet. Press information 98-6, 2 September 1998. Centre for Plant Breeding and Reproduction Research (CPRO-DLO), Wageningen, the Netherlands.
Crabb, A.R. (1947) *The Hybrid-Corn Makers: Prophets of Plenty.* Rutgers University Press, New Brunswick, United States.
CSSA (1964) *Plant Breeders' Rights.* ASA Special Publication 3, March 1964, the Crop Science Society of America, Madison, United States.
Cueto, M. (ed.) (1994) *Missionaries of Science: The Rockefeller Foundation in Latin America.* Indiana University Press, Bloomington and Indianapolis, United States.
Cunningham, I.S. (1992) William Lacy Brown 1913-1991: scientist, executive and mentor. *Diversity* 8(2), 15-22.
Curagen (1998) Curagen Internet homepage. (http://www.curagen.com) (consulted December 1998).
De Haan, H. (1940) De rassenlijst voor landbouwgewassen: overzicht 1924-1939. [The list of races for agricultural crops. Survey 1924 -1939.] Offprint from Jaarboek 1940 Algemeene Bond van Oud-leerlingen van Inrichtingen voor Middelbaar Landbouwonderwijs. Wageningen, the Netherlands. (summary in English)
De Haan, H. (1949) Vijf en twintig jaren rassenlijst. [Twenty-five years of the Dutch list of varieties]. Wageningen, the Netherlands. (summary in English)
De Souza Silva, J. (1989) Science and the changing nature of the struggle over plant genetic resources: from plant hunters to plant crafters. PhD thesis, The University of Kentucky, Lexington, United States.
De Vroey, M. (1984) A regulation approach interpretation of contemporary crisis. *Capital and Class* 23, 44-66.
DeKalb (1998) DeKalb Internet homepage. (http://www.dekalb.com) (consulted January 1999).
Del Pilar Valencia, M. (1996) Fundamentos juridicos de los derechos intellectuales colectivos. Paper presented at the Seminario Gestion de la Propiedad Intelectual en Biotecnología, May 6-11 1996, CIAT, Cali, Colombia. Organizaci6n Indígena de Antioqiua (OIA), Medellín, Colombia.
Dixon, B. (1993) Mendel's heirs and the summer of 1939. *Bio/Technology* 11, 646.
Dobb, M. (1972) *Soviet Economic Development Since 1917.* Routledge & Kegan Paul Ltd, London, United Kingdom.
Doherty, A. (1994) The role of non-governmental organizations in UNCED. In: Spector *et al.* (1994), pp. 199-218.
Dosi, G., Freeman, C., Nelson, R., Silverberg G., Soete, L. (eds.) (1988) *Technological Change and Economic Theory.* Pinter Publishers, London, United Kindom and New York, United States.
Dowswell, C.R., Paliwal, R.L., Cantrell, R.P. (1996) *Maize in the Third World.* Westview Press, Boulder, United States.
Doyle, J. (1985) *Altered Harvest. Agriculture, Genetics, and the Fate of the World's Food*

Supply. Viking Penguin, New York, United States.

DuPont (1999) DuPont Internet homepage. (http://www.dupont.com) (consulted January 1999).

Duvick, D.N (1977) Genetic rates of gains in hybrid maize yields during the past 40 years. *Maydica* 21, 187-196.

Duvick, D.N. (1984a) Genetic contributions to yield gains of U.S. hybrid maize, 1930 to 1980. In: Fehr (ed.) (1984), pp. 15-47.

Duvick, D.N. (1984b) Genetic diversity in major farm crops on the farm and in reserve. *Economic Botany* 38, 161-178.

Duvick, D.N. (1996) Utilization of biotechnology in US plant breeding. *Biotechnology and Development Monitor* 27, 15-17.

Duvick, D.N. and Shands, H. (1994) The national genetic resources program and its advisory council: a discussion by Henry Shands and Donald Duvick. *Diversity* 10(4), 51-55.

Echeverría, R., Trigo, E., Byerlee, D. (1996) Institutional change and effective financing of agricultural research in Latin America. Findings of a workshop organized by the Inter-American Development Bank and the World Bank in Buenos Aires, August 1995. World Bank Technical Paper 330, World Bank, Washington D.C., United States.

EuropaBio (1998) Internet homepage of the European Association for Bio-Industries, Brussels, Belgium. (http://www.europa-bio.be) (consulted January 1999).

FAO (1950a) *World Catalogue of Genetic Stocks: Rice.* Food and Agriculture Organization, Rome, Italy.

FAO (1950b) *World Catalogue of Genetic Stocks: Wheat.* Food and Agriculture Organization, Rome, Italy.

FAO (1957-1970) *FAO Plant Introduction Newsletter.* Food and Agriculture Organization, Rome, Italy. (various issues).

FAO (1958a) *Tabulated Information on Tropical and Subtropical Grain Legumes*. Food and Agriculture Organization, Rome, Italy.

FAO (1958b) *The World Seed Campaign News* 1. Food and Agriculture Organization, Rome, Italy.

FAO (1959a) *The World Seed Campaign News* 2. Food and Agriculture Organization, Rome, Italy.

FAO (1959b) *World Catalogue of Genetic Stocks: Barley.* Food and Agriculture Organization, Rome, Italy.

FAO (1960a) *The World Seed Campaign News* 7. Food and Agriculture Organization, Rome, Italy.

FAO (1960b) *The World Seed Campaign News* 9. Food and Agriculture Organization, Rome, Italy.

FAO (1962) *FAO Feature.* December issue. Food and Agriculture Organization, Rome, Italy.

FAO (1969) Report of the third session of the FAO Panel of Experts on Plant Exploration and Introduction. Food and Agriculture Organization, Rome, Italy.

FAO (1973) Fifth session of the FAO Panel of Experts on Plant Exploration and Introduction. Food and Agriculture Organization, Rome, Italy.

FAO (1985) Report of the first session of the Commission on Plant Genetic Resources. March 1985. Food and Agriculture Organization, Rome, Italy.

FAOSTAT (1998) FAOSTAT statistics database. Food and Agriculture Organization, Rome, Italy. (http://www.fao.org/cgi-bin/nph-db.pl?/subset=agriculture) (consulted October 1998).

Fehr, W.R. (ed.) (1984) Genetic Contributions to Yield Gain in Five Major Crop Plants.

Proceedings of a symposium of the Crop Science Society of America. American Society of Agronomy, Madison, United States.

Fitzgerald, D. (1994) Exporting American agriculture: the Rockefeller Foundation in Mexico 1943-1953. In: Cueto (ed.) (1994), pp. 72-96.

Fitzpatrick, K. and Scarth, R. (1998) Improving the health and nutritional value of oilseeds. *Plant Biotechnology Institute Bulletin* January.

Flitner, M. (1995) *Sammler, Räuber und Gelehrte: Die Politischen Interessen an Pflanzengenetischen Ressourcen 1895-1995.* Campus Verlag, Frankfurt, Germany.

Flórez, M. (1996) Son las regulaciones coexistentes y consistentes entre si? Paper presented at the Seminario Gestion de la Propiedad Intelectual en Biotecnología, May 6-11 1996, CIAT, Cali, Colombia. Instituto Latinoamericano de Sevicios Legales Alternativos (ILSA), Bogotá, Colombia.

Flynn, J., Carey, J., and Echikson, W. (1998) The pros and cons of gene-spliced food: Biotech companies are facing a growing furor in Europe over genetically altered food. *Business Week* February 2, 1998.

Fowler, C. and Mooney, P. (1990) *Shattering: Food Politics, and the Loss of Genetic Diversity.* University of Arizona Press, Tucson, United States.

Fowler, C. (1994) *Unnatural Selection: Technology, Politics, and Plant Evolution.* Gordon and Breach, Yverdon, Switzerland.

Fowler, C., Lachkovics, E., Mooney, P., Shand, H. (1988) The laws of life: another development and the new biotechnologies. *Development Dialogue* 1-2, Dag Hammerskjöld Foundation, Uppsala, Sweden.

Francis, D. (1998) Bioscience will boom next century. *The Financial Post,* 21 February.

Frankel, O.H. and Soulé, M.E. (1981) *Conservation and Evolution.* Cambridge University Press, Cambridge, United Kingdom.

Freiberg, B. (1997a) Groupe Limagrain: diversifying into scientific opportunities. *Seed and Crops Digest* 48(5), June/July, 8-10,21,49.

Freiberg, B. (1997b) New technology: lots of promise....but not an easy ride. *Seed and Crops Digest* 48(8), November, 4-6.

Freiberg, B. (1997c) DeKalb. *Seed and Crops Digest* 48(5), June/July, 4-6,55,60.

Friedland, W., Busch, L., Buttel, F., Rudy, A. (1991) *Towards a New Political Economy of Agriculture.* Westview Press, Boulder, United States.

Friedland, W. (1994) The global fresh fruit and vegetable system: an industrial organization analysis. In: McMichael, Ph. (ed.) *The Global Restructuring of Agro-Food Systems.* Cornell University Press, Ithaca, United States, and London, United Kingdom, pp. 173-189.

Friedmann, H. (1992) Distance and durability: shaky foundations of the world food economy. *Third World Quarterly* 13(2), 371-383.

Friedmann, H. and McMichael, Ph. (1989) Agriculture and the state system. The rise and decline of national agricultures, 1870 to the present. *Sociologia Ruralis* 24(2), 93-117.

GATT (1990) Communication from Argentina, Brazil, Chile, China, Colombia, Cuba, Egypt, India, Nigeria, Peru, Tanzania, and Uruguay. Negotiating Group on Trade-Related Aspects of Intellectual Property Rights, including Trade in Counterfeit Goods. Secretariat of the General Agreement on Tariffs and Trade, Geneva, Switzerland.

Gonzales, C. and Jaramillo, C.F. (eds.) (1994) *Competitividad Sin Probreza. Estudios Para el Desarrollo Del Campo en Colombia.* Fonade TM Editores, Bogotá, Colombia.

Goodman, D., Sorj, B., Wilkinson, J. (1987) *From Farming to Biotechnology. A Theory of Agro-Industrial Development.* Basil Blackwell, Oxford, United Kingdom.

Gourevitch, P. (1986) *Politics in Hard Times. Comparative Responses to International Economic Crises.* Cornell University Press, Ithaca, United States, and London, United Kingdom.

GRAIN (1998) GEF: an unsuitable vehicle for biodiversity conservation. Position paper published by the Genetic Resources Action International. Barcelona, Spain. (http://www.grain.org/publications/reports).

Grove, R.H. (1995) *Green Imperialism: Colonial Expansion, Tropical Island Edens and the Origins of Environmentalism, 1600-1860.* Cambridge University Press, Cambridge, United Kingdom.

Hamilton, N.D. (1993) Who owns dinner: evolving legal mechanisms for ownership of plant genetic resources, or will recognizing intellectual property rights in plants reshape international agriculture? White Paper 93-1. Agricultural Law Center, Drake University, Des Moines, United States.

Hamilton, N.D. (1996) Possible effects of recent developments in plant related intellectual property rights in the U.S.. In: Van Wijk and Jaffé (eds.) (1996), pp. 84-99.

Hammer, K. (1993) The 50th aniversary of the Gatersleben genebank. *FAO/IPBGR Plant Genetic Resources Newsletter* 91/92, 1-8.

Hardin, G. (1968) The tragedy of the commons. *Science* 162, 1243-1248.

Hardon, J. (1988) Crop genetic resources conservation in Western Europe. Centre for Genetic Resources Netherlands, Wageningen, the Netherlands.

Harlan, J. and Martini, M.L. (1936) Problems and results in barley breeding. In: *Yearbook of Agriculture 1936.* USGPO, Washington, D.C., United States.

Harmston, R. (1979) Foreword. In: Mooney (1979), pp. vi-vii.

Hawkes, J.G. and Lamberts, H. (1976) EUCARPIA's fifteen years of activities in genetic resources. *Euphytica* 26, 1-3.

Heijbroek, A.M.A., Schutter, E.M.L., Van Gaasbeek, A.F. (1996) *The World Seed Market. Development and Strategies.* Second revised edition. Rabobank, Utrecht, the Netherlands.

Heitz (1987) The history of plant variety protection. In: UPOV, *The First Twenty-Five Years of the International Convention for the Protection of New Varieties of Plants.* International Union for the Protection of New Varieties of Plants (UPOV), Geneva, Switzerland, pp. 53-96.

Heitz, A. (1991) The history of the UPOV convention and the rationale for plant breeder's rights. In: UPOV, *Seminar on the nature of and rationale for the protection of plant varieties under the UPOV Convention.* Budapest, September 19-21, 1990. International Union for the Protection of New Varieties of Plants, Geneva, Switzerland, pp. 19-42.

Hermitte, M.A. (1988) Histoire juridiques extravagantes. La reproduction végétale. In: Edelman, B. and Hermitte, M.A. (eds.), *L'homme, La Nature, et Le Droit.* Christian Bourgois, Paris, France, pp. 40-82.

Hobbelink, H. (1987) *New Hope or False Promise? Biotechnology and Third World Agriculture.* International Coalition for Development Action, Brussels, Belgium.

Hodgson, R.E. (ed.) (1961) *Germ Plasm Resources.* American Association for the Advancement of Science, Washington, D.C., United States.

Hopkins, R.F. and Puchala, D.J. (eds.) (1978) The global political economy of food. *International Organization* 32(3).

Hougas, R.W. (1984) Plant germplasm policy. In: Yeatman *et al.* (1984), pp. 15-29.

Hudson, P.S. and Richens, R.H. (1946) *The New Genetics in the Soviet Union.* Imperial Bureau of Plant Breeding and Genetics, Cambridge, United Kingdom.

Hyams, E. (1971) *Plants in the Service of Man, 10,000 years of Domestication.* Dent & Sons, London, United Kingdom.
Hyland, H.L. (1963) Plant introduction objectives and procedures in the United States. *Genetica Agraria* 17(1-4), 470-482.
Hyland, H.L. (1984) History of plant introduction in the United States. In: Yeatman *et al.* (eds.) (1984), pp. 1-14.
IAN (1998) Linking the Food Chain. Public statement by the International Agri-Food Network. (http://www.agrifood.net/linking) (consulted January 1999).
IBPGR (1991) *Elsevier's Dictionary of Plant Genetic Resources.* Elsevier, Amsterdam, the Netherlands.
ICC (1998) Business and the global economy. ICC statement on behalf of world business to the Heads of State and Government attending the Birmingham Summit, 15-17 April, 1998. International Chamber of Commerce. (http://www.iccwbo.org/commissions).
IPC, KEIDANREN, UNICE (1988) Basic framework of GATT provisions on intellectual property. Statement of views of the European, Japanese, and United States business communities. Intellectual Property Committee (IPC), USA; Japan Federation of Economic Organizations (KEIDANREN), Japan; Union of Industrial and Employers' Confederations of Europe (UNICE), Europe. June 1988.
IPGRI (1998) IPGRI Internet homepage. International Plant Genetic Resources Institute, Rome, Italy. (http://www.cgiar.org.ipgri/cgsuppor/index.htm) (consulted December 1998).
Jaffé, W. and Van Wijk, J. (1995) The impact of plant breeders' rights in developing countries: debate and experience in Argentina, Chile, Colombia, Mexico and Uruguay. Inter-American Institute for Cooperation on Agriculture (IICA), Costa Rica, and Directorate General International Cooperation, Ministry of Foreign Affairs, the Netherlands.
James, C. (1997) Progressing public-private sector partnerships in international agricultural research and development. ISAAA Briefs 4. International Service for the Acquisition of Agrobiotech Applications, Ithaca, United States.
James, E. (1961) Perpetuation and protection of germ plasm as seed. In: Hodgson (1961), pp. 317-326.
Jones, Q. and Wolff, I.A. (1961) Using germ plasm for new products. In: Hodgson (1961), pp. 265-277.
Juma, C. (1989) *The Gene Hunters: Biotechnology and the Scramble for Seeds.* Zed Books, London, United Kingdom.
Junne, G. (ed.) (1987) Biotechnology - The will to manipulate is human. But what do we want? *Development: Seeds of Change.* Journal of the Society For International Development 4.
KAW (1923) A short history and the present position of botanical investigation in the Dutch East Indies. Koninklijke Academie van Wetenschappen, Amsterdam, the Netherlands.
Kay, C. (1996) The agrarian question in Chile since the democratic transition. Working Paper Series No.219, Institute of Social Studies (ISS), The Hague, the Netherlands.
Kenney, M., Lobao, L., Curry, J., Goe, W. (1991) Agriculture in U.S. Fordism: the integration of the productive consumer. In: Friedland *et al.* (1991), pp. 173-188.
Keohane, R.O and Nye, J.S. (1977) *Power and Interdependence: World Politics in Transition.* Little, Brown and Company, Boston, United States.
Kindleberger, C.P. (1973) *The World Depression 1929-1939.* Allen Lane the Penguin Press, London, United Kingdom.
Kjaer, A. (1955) The desirability of a uniform international terminology in relation to seed

certification schemes. In: OEEC (1955), pp. 83-90.
Kloppenburg Jr, J.R. (1988a) *First the Seed. The Political Economy of Plant Biotechnology 1492-2000.* Cambridge University Press, New York, United States.
Kloppenburg Jr, J.R. (ed.) (1988b) *Seeds and Sovereignty: The Use and Control of Plant Genetic Resources.* Duke University Press, Durham, United States.
Kloppenburg, J.R. and Kleinman, D.L. (1988) Seeds of controversy: national property versus common heritage. In: Kloppenburg Jr. (ed.) (1988b), pp. 173-203.
Knudson, M. and Ruttan, V. (1988) Research and development of a biological innovation: commercial hybrid wheat. Food Research Institute Studies 21(1), 45-68.
Kollek, R. (1995) Ambiguous genes. *Biotechnology and Development Monitor* 23, 24.
Koning, N. (1986) Agrarische gezinsbedrijven en industrieel kapitalisme. *Tijdschrift voor Politieke Economie* 6(1), 35-67.
Koning, N. (1994) *The Failure of Agrarian Capitalism: Agrarian Politics in the UK, Germany, the Netherlands and the USA, 1846-1919.* Routledge, London, United Kingdom, and New York, United States.
Krasner, S.D. (1982) Structural causes and regime consequences: regimes as intervening variables. *International Organization* 36(2), 185-206.
Krasner, S.D. (1985) *Structural Conflict: The Third World Against Global Liberalism.* University of California Press, Berkeley, United States.
Krol, J. (1997) Sustainable growth - a business imperative. Speech of DuPont president Jack Krol to the Environmental Northern Seas Conference, August 26, 1997 in Stavanger, Norway. (http://www.dupont.com/corp/whats-new/speeches/krol).
Laclavière, B. (1972) The future of ASSINSEL. *Bulletin ASSINSEL* 4, December, 54-55.
Laird, R. and Laird, B. (1970) *Soviet Communism and Agrarian Revolution.* Penguin Books, Hamondsworth, United Kingdom.
Lange, P. (1985) The nature of plant breeders' rights (plant variety protection law) and their demarcation from patentable inventions. In: UPOV, Industrial Patents and Plant Breeders' Rights - Their Proper Fields and Possibilities for Their Demarcation. *Records of a Symposium held on the occasion of the eighteenth ordinary session of the Council of the International Union for the Protection of New Varieties of Plants (UPOV), Geneva, October 17, 1984.* UPOV, Geneva, Switzerland, pp. 29-40.
Lange, P. (1997) The non-patentability of plant varieties. The decision of the Technical Board of Appeal 3.3.4 of February 21, 1995-T356/93. *Plant Variety Protection.* Gazette and Newsletter of the International Union for the Protection of New Varieties of Plants (UPOV) 83, 25-33.
Le Buanec, B. (1996) Globalization of the seed industry: current situation and evolution. *Seed Science & Technology* 24, 409-417.
Leskien, D. and Flitner, M. (1997) Intellectual property rights and plant genetic resources: options for a *sui generis* system. Issues in Genetic Resources 6. International Plant Genetic Resources Institute (IPGRI), Rome, Italy.
Lewontin, R.C. and Berlan, J.P. (1990) The political economy of agricultural research: the case of hybrid corn. In: Caroll, C.R., Vandermeer, J,H., Rosset, P., *Agroecology.* McGraw-Hill Publishing Company, Washington, D.C., United States, pp. 613-628.
Lewontin, S. (1983) The Green Revolution and the politics of agricultural development in Mexico since 1940. PhD thesis. UMI Dissertation Service, Ann Arbor, United States.
Limagrain (1997) Groupe Limagrain 1996/1997. Annual report. Limagrain, Shappes, France.
Limagrain (1999) Limagrain Internet homepage. Limagrain, France.

(http://www/limagrain.com) (consulted January 1999).
Loskutov, I. *Vavilov and his Institute.* International Plant Genetic Resources Institute (IPGRI), Rome, Italy. (in press)
Machlup, F. (1958) An economic review of the patent system. Subcommittee on Patents, Trademarks, and Copyrights of the Committee on the Judiciary. U.S. Senate Study 15. US Government Printing Office, Washington D.C., United States.
Machlup, F and Penrose, E. (1950) The patent controversy in the nineteenth century. *The Journal of Economic History* 10(1), 1-29.
MAFF (1999) Japanese Ministry of Agriculture, Forestry and Fisheries genome research project. Ministry of Agriculture, Forestry and Fisheries, Tokyo, Japan. (http://www.bank.dna.affrc.go.jp) (consulted January 1999).
Manzur, M.I. (1995a) Derechos de propriedad intelectual. Serie Documentos, Comité Nacional Pro Defensa de la Fauna y Flora (CODEFF), Santiago, Chile.
Manzur, M.I. (1995b) La Posición de CODEFF en relación a la ley Chilena de obtenerse de nuevas variedades végétales (Ley No.19.342). Comité Nacional Pro Defensa de la Fauna y Flora (CODEFF), Santiago, Chile.
Mayr, E. (1982) *The Growth of Biological Thought: Diversity, Evolution and Inheritance.* Belknap Harvard, London, United Kingdom.
Masefield, G.B. (1972) *A History of the Colonial Agricultural Service.* Clarendon Press, Oxford, United Kingdom.
Massieu, Y. (1998) ELM: a new global player in the vegetable market. *Biotechnology and Development Monitor* 34, 9-13.
McCarthy, J.Th. (1995) Intellectual property - America's overlooked export. *University of Dayton Law Review* 20(3), 809-819.
McClellan, S. (1991) Theorizing new deal farm policy: broad constraints of capital accumulation and the creation of a hegemonic relation. In: Friedland *et al.* (1991), pp. 215-231.
McCormick, J. (1989) *The Global Environmental Movement.* Bellhaven Press, London, United Kingdom.
McMullen, N. (1987) *Seeds and World Agricultural Progress.* National Planning Association, Washington, D.C., United States.
Millennium (1998) Millennium Internet homepage. (http://www.millennium.com) (consulted January 1999).
Ministerio de Agricultura (1994) Ley de Obtentores de Nuevas Variedades Végétales No. 19.342. Santiago, Chile.
Ministerio de Economía, Fomento y Reconstrucción (1991) Ley No.19.039 Establece normas aplicables a los privilegios industriales y protección de los derechos de propiedad industrial. Santiago, Chile.
Monaco, L.C. (1977) Consequences of the introduction of coffee rust into Brazil. In: Day, P. (ed.) *Genetic Basis of Epidemics in Agriculture.* The New York Academy of Sciences, New York, United States, pp. 56-71.
Monsanto (1998a) Monsanto/IBM strategic technology alliance. Press release. (http://www.monsanto.com/monsanto/media/).
Monsanto (1998b) Monsanto Internet homepage. (http://www.monsanto.com/monsanto/financial/quarterly) (consulted January 1999).
Montecinos, C. (1996) *Sui generis* - A dead end alley. *Seedling* 13(4), 19-28.
Mooney, P.R. (1979) *Seeds of the Earth. A Private or Public Resource?* Canadian Council for International Co-operation and the International Coalition for Development Action

(ICDA), Ottawa, Canada, and London, United Kingdom.
Mooney, P.R. (1983) The law of the seed. Another development and plant genetic resources. *Development Dialogue* 1-2, Dag Hammerskjöld Foundation, Uppsala, Sweden.
Mooney, P.R. (1998) The parts of life: agricultural biodiversity, indigenous knowledge, and the role of the third system. *Development Dialogue* (special issue 1998) Dag Hammarskjöld Foundation, Uppsala, Sweden.
Muchnik, E. (1996) Las reformas econ6micas y la agricultura en Chile. In: Olavarría and Rojas (1996), pp. 117-125.
Mugabe, J., Barber, C., Henne, G., Glowka L., La Viña, A. (eds.) (1997) *Access to Genetic Resources. Strategies for Sharing Benefits.* ACTS Press, Nairobi, Kenya.
Nap, J.P.H. (1999) Plant Biotechnology: the quest for useful pieces of DNA. In: Almekinders, C.J.M. and De Boef, W.S. (eds.) *Towards a Synthesis between Crop Conservation and Development.* Intermediate Technologies, London, United Kingdom. (in press)
NCGI (1998) NCGI Internet homepage. National Corn Genome Initiative. (http://www.inverizon.com/ncgi) (consulted December 1998).
NCGR (1998) NCGR Internet home page. National Center for Genome Resources (http://www.ncgr.org) (consulted December 1998).
Neville-Rolfe, E. (1984) *The Politics of Agriculture in the European Community.* Policy Studies Institute, London, United Kingdom.
Newman, M., Fulton T., Glaser, L. (1987) A comparison of agriculture in the United States and the European community. Foreign Agriculture Economic Report 233. Economic Research Service, U.S. Department of Agriculture, Washington, D.C., United States.
Nilsson-Leissner, G. (1955) Present-day status of seed production, testing and distribution in the OEEC participating countries. In: OEEC (1955), pp. 25-36.
Novartis (1998a) Novartis Internet homepage. (http://www.novartis.com) (consulted January 1999).
Novartis (1998b) The world of Novartis Seeds. Novartis, Bern, Switzerland.
NTZ (1992) *Twee Eeuwen Tuinbouwzaden.* Nederlandse Vereniging voor het Tuinzaadbedrijfsleven, The Hague, the Netherlands.
Oasa, E.K. (1981) The International Rice Research Institute and the Green Revolution: A case study on the politics of agricultural research. PhD thesis. UMI Dissertation Service, Ann Arbor, Michigan, United States.
ODEPA (1998) Internet homepage of the Oficina de Estudios y Politicas Agrarias, Ministerio de Agricultura, Santiago, Chile. (http://www.minagri.gob.cl/minagri/serinfo/datos.html) (consulted December 1998).
OEEC (1955) *Development of Seed Production and the Seed Trade in Europe.* European Productivity Agency of the Organization of European Economic Co-operation, Paris, France.
Olavarría, J. and Rojas, A. (eds.) (1996) *Implicancias del NAFTA en la Agricultura Chilena.* Editorial Universidad de Talca, Talca, Chile.
Oman, R. (1994) Intellectual property - our once and future strength. *The George Washington Journal of International Law and Economics* 27(2,3), 301-303.
Orsenigo, L. (1989) *The Emergence of Biotechnology. Institutions and Markets in Industrial Innovation.* Pinter Publishers, London, United Kingdom.
Osborne, R. (1997) Monsanto to pay $218 m. in pact with Millennium. *BioWorld Today* 8(210), 1-3.
Palladino, P. (1993) Between craft and science: plant breeding, Mendelian genetics, and

British universities, 1900-1920. *Technology and Culture* 300-323.
Palladino, P. (1996) Science, technology, and the economy: plant breeding in Great Britain, 1920-1970. *Economic History Review* 69(1), 116-136.
Pardey, P.G., Alston, A.M., Christian, J.E., Fan, S. (1996) *Hidden harvest: U.S. benefits from international research aid.* Food Policy Report. International Food Policy Research Institute (IFPRI), Washington, D.C., United States.
Paul, D.B and Kimmelman, B.A. (1988) Mendel in America: theory and practice, 1900-1919. In: Rainger, R., Benson K.R., Maienschein J. (eds.) *The American Development of Biology.* Pennsylvania Press, Pennsylvania, United States, pp. 281-310.
Paul, H.W. (1985) *From Knowledge to Power: The Rise of the Science Empire in France, 1860-1939.* Cambridge University Press, Cambridge, United Kingdom.
Pearce, D. and Moran, D. (1994) *The Economic Value of Biodiversity.* IUCN and Earthscan, London, United Kingdom.
Perkins, J.H. (1997) *Geopolitics and the Green Revolution: Wheat, Genes, and the Cold War.* Oxford University Press, Oxford and New York, United Kingdom and United States.
Perlas, N. and Vellvé, R. (1995) *Oryza Nirvana? An NGO Review of the International Rice Research Institute in Southeast Asia.* South East Asia Regional Institute for Community Development (SEARICE), Manila, the Philippines.
Pioneer Hi-Bred (1998) 1997 Annual report of Pioneer Hi-Bred. Pioneer Hi-Bred, Des Moines, United States.
Pioneer Hi-Bred (1999) Pioneer Internet homepage. (http://www.pioneer.com) (consulted December 1998).
Pistorius, R. (1995a) Forum shopping: issue linkages in the genetic resources issue. In: Bartlett, R.V., Kurian, P.A., Malik, M. (eds.) *International Organizations and Environmental Policy.* Greenwood Press, London, United Kingdom, pp. 209-222.
Pistorius, R. (1995b) Will material transfer agreements open Pandora's box? *Biotechnology and Development Monitor* 25, 20-22.
Pistorius, R. (1997) *Scientists, Plants and Politics: A History of the Plant Genetic Resources Movement.* International Plant Genetic Resources Institute (IPGRI), Rome, Italy.
Pistorius, R. and Van Wijk, J. (1993) Biodiversity prospecting: commercializing genetic resources for export. *Biotechnology and Development Monitor* 15, 12-15.
Plucknett, D.L., Smith, N.J.H., Williams, J.T., Anishetti, N.M. (1987) Crop germplasm conservation and developing countries. *Science* 220, 163-169.
Pombo, D. (1996) Derechos de los agricultores en el marco de la discusión sobre recursos fitogenéticos. Paper presented at the Seminario Gestion de la Propiedad Intelectual en Biotecnología, May 6-11 1996, CIAT, Cali, Colombia. Instituto de Gestión Ambiental, Bogotá, Colombia.
Pouillet, M. and Plé, M. (1896) *La Convention D'Union de Paris du 20 Mars 1883 pour la Protection de la Propriété Industrielle: Commentaire.* Marchal et Billard, Paris, France.
Powledge, F. (1994) Genetic resources issues dominate CGIAR agenda: more challenges and less money for beleaguered group. *Diversity* 10(1), 4-8.
Pratt, E.T. (1995) Remarks of Edmund T. Pratt, Jr, Chairman Emiritus, Pfizer Inc., at the USCIB Conference on Intellectual Property. Pfizer, Washington, D.C., United States.
Puchala, D.J. and Hopkins, R.F. (1982) International regimes: lessons from inductive analysis. *International Organization* 36(2), 245-275.
Quiroz, J. (1996) La Agricultura y agroindustria nacionales frente al NAFTA, In: Olavarría and Rojas (eds.) (1996), pp. 127-129.
RAFI (1988) A report on germplasm embargoes. *RAFI Communique* October 1988. Rural

Advancement Fund International, Durham, Canada.

RAFI (1995a) Utility plant patents: A review of the US experience (1985 - July 1995). *RAFI Communique* July/August 1995, Rural Advancement Foundation International, Ottawa, Canada.

RAFI (1995b) Indigenous person from Papua New Guinea claimed in U.S. government patent. Press release of the Rural Advancement Foundation International, Ottawa, Canada, October 1995.

RAFI (1998a) Seed industry consolidation: who owns whom? *RAFI Communique* July/August 1998, Rural Advancement Foundation International, Osborne, Canada.

RAFI (1998b) The terminator technology: new genetic technolology aims to prevent farmers from saving seed. *RAFI Communique* March/April 1998, Rural Advancement Foundation International, Osborne, Canada.

RAFI (1998c) And now, the "Verminator"! Fat cat corp. with fat rat gene can kill crops. Rural Advancement Foundation International. (http://www.rafi.ca).

RAFI (1998d) Report of the CGIAR External Review Panel. RAFI Translator. Rural Advancement Foundation International. (http://rafi.org/translator/).

Rangekar, D. (1998) Innovative appropriation - transforming plants into products: Wheat breeding in the context of the Plant Varieties & Seeds Act (1964). Draft Ph.D. thesis, School of Economics, Kingston University, Kingston, United Kingdom.

Reid, W.V., Laird, S.A., Meyer, C.A., Gaméz, R., Sittenfeld, A., Janzen, D.H., Gollin, M.A., Juma, C. (1995) *Biodiversity Prospecting: Using Genetic Resources for Sustainable Development.* World Resources Institute, Washington, D.C., United States.

Rejesus, R.M., Smale, M., Van Ginkel, M. (1996) Wheat breeders' perspective on genetic diversity and germplasm use: findings from an international survey. *Plant Varieties and Seeds* 9, 129-147.

República de Colombia (1996) Sistema Nacional Ambiental. Ley 99 de diciembre 1993.

Rittberger, V. (1988) International regimes and peaceful conflict regulation. In: Wallensteen, P. (ed.) *Peace Research.* Westview Press, Boulder, United States, pp. 144-165.

Robinson, R.A. (1996) *Return to Resistance: Breeding Crops to Reduce Pesticide Dependence.* AgAccess, Davis, California, United States.

Roobeek, A.J.M. (1990), *Beyond the Technology Race. An Analysis of Technology Policy in Seven Industrialized Countries.* Elsevier, Amsterdam, the Netherlands.

Rose, G. (ed.) (1993) Transfer of biotechnology and genetic resources. Special edition of the *Review of European Community & International Environmental Law* 2(2).

Ruff, A.M. and Worthington, L.F. (1998) The Convention on Biological Diversity: the "mother of all conventions". *Diversity* 14(1&2), 6-8.

Ruggie, J.G. (1975) International responses to technology: concepts and trends. *International Organization* 29, 557-584.

Ruggiero, R. (1998) A global system for the next fifty years. Speech by WTO Director-General Ruggiero on October 30 1998, London, United Kingdom. World Trade Organization, Geneva, Switzerland. (http://www.wto.org/wto/speeches).

Ruivenkamp, G. (1989) *De Invoering van Biotechnologie in de Agro-Industriële Productieketen. De Overgang naar een Nieuwe Arbeidsorganisatie.* Jan van Arkel, Utrecht, the Netherlands.

Saloutos (1982) *The American Farmer and the New Deal.* The Iowa State University Press, Ames, Iowa, United States.

Sánchez, V. and Juma, C. (eds.) (1994) *Biodiplomacy. Genetic Resources and International Relations.* ACTS Press, Nairobi, Kenya.

Schiff, E. (1971) *Industrialization without National Patents. The Netherlands, 1869-1912; Switzerland, 1850-1907.* Princeton University Press, Princeton, United States.
Schmidt, J.W. (1984) Genetic contributions to yield gains in wheat. In: Fehr (ed.) (1984), pp. 89-101.
Seghal, S. (1996) IPR driven restructuring of the seed industry. *Biotechnology and Development Monitor* 29, 18-21.
Sell, S.K. (1995) The origins of a trade-based approach to intellectual property protection. *Science Communication* 17(2), 163-185.
Siebeck, E.W. (ed.) (1990) Strengthening protection of intellectual property in developing countries: a survey of the literature. World Bank Discussion Paper 112. The World Bank, Washington D.C., United States.
Sirks, M.J. (ed.) (1936) International congress of plant breeders. *Reports presented to the congress on the general organization of plant breeding in various countries, the present problems of wheat breeding, of potato breeding and of sugar beet breeding.* Brill, Leiden, the Netherlands.
Sneep, J., Hendriksen, A.J.T., Holbek, O. (eds.) (1979) *Plant Breeding Perspectives: Centennial Publication of Koninklijke Kweekbedrijf en Zaadhandel D.J. van der Have.* Pudoc, Wageningen, the Netherlands.
Somani, L.L. (1992) *Dictionary of Plant Breeding and Seed Technology.* Agrotech Publishing Academy, Udaipur, India.
Sommerville, C. (1997) Secrecy of agricultural genome databases criticized. *Science* 27 June, 2.
Spector, B.I., Sjöstedt G., Zartman, I.W. (1994) *Negotiating International Regimes: Lessons Learned from the United Nations Conference on Environment and Development (UNCED).* Graham & Trotman and Martinus Nijhoff, Laxenburg, Austria, and London, United Kingdom.
Srivastava, J., Smith, N.J.H., Forno, D. (1996) Biodiversity and agriculture: implications for conservation and development. World Bank Paper 321. World Bank, Washington, D.C., United States.
Stallman, J.I. and Schmid, A.A. (1987) Property rights in plants: implications for biotechnology research and extension. *American Journal of Agricultural Economics* 69(2), 432-437.
Strauch, M. (1972) The Plant Variety Protection Act of the United States and the Convention of Paris 1961. Reflections on parallels and differences. *Bulletin Assinsel* 4, December 1972, pp. 31-37.
Strauss, D. (1995) New genetic resources policy unit to assist CGIAR avoid 'minefields'. *Diversity* 11(3), 2.
Stubbe, H. (1971) *History of Genetics: From Prehistoric Times to the Rediscovery of Mendel's Laws.* Massachusetts Institute of Technology, Massachusetts, United States.
Swanson, T.M., Pearce, D.W., Cervigni, R. (1994) The appropriation of the benefits of plant genetic resources for agriculture: an economic analysis of the alternative mechanisms for biodiversity conservation. FAO Background Study 1. Food and Agriculture Organization, Rome, Italy.
Talbot, R.B. (1990) *The Four World Food Agencies in Rome.* Iowa State University Press, Ames, United States.
Tanksley, S.D. and McCouch, S.R. (1997) Seed banks and molecular maps: unlocking genetic potential from the wild. *Science* 277(22), 1063-1066.
Thielebein, M. (1955) Possibilities of seed production and multiplication in the most

suitable areas of the participating countries. In: OEEC (1955), pp. 103-110.
Thirtle C.G., Piesse, J., Smith V.H. (1997) An economic approach to the structure, historical development and reform of agricultural R&D in the UK. CAS Paper 35. Centre for Agricultural Strategy, United Kingdom.
Thrupp, L.A., Bergeron, G., Waters, W. (1995) *Bittersweet Harvests for Global Supermarkets: Challenges in Latin America's Agricultural Export Boom.* World Resources Institute, Washington, D.C., United States.
Tracy, M. (1982) *Agriculture in Western Europe. Challenge and Response 1880-1980,* 2nd edn. Granada Press, London, United Kingdom.
Tripp, R. (ed.) (1997) *New Seed and Old Laws. Regulatory Reform and the Diversification of National Seed Systems.* Intermediate Technology Publications on behalf of the Overseas Development Institute, London, United Kingdom.
UNEP (1992) Convention on Biological Diversity. United Nations Environmental Programme, Nairobi, Kenya.
United Nations (1975) The role of the patent system in the transfer of technology to developing countries. Report prepared jointly by the United Nations Department of Economic and Social Affairs, the UNCTAD secretariat and the International Bureau of the WIPO. United Nations, New York, United States.
UPOV (1991), International Convention for the Protection of New Varieties of Plants of December 2, 1961, as Revised at Geneva on November 10, 1972, on October 23, 1978, and on March 19, 1991. International Union for the Protection of New Varieties of Plants, Geneva, Switzerland.
UPOV (1998) Colombia. *Plant Variety Protection.* Gazette and Newsletter of the International Union for the Protection of New Varieties of Plants 84, 33.
USDA (1995) AREI updates: agricultural research. Updates on Agricultural Resources and Environment Indicators 5 (revised), Economic Research Service, U.S. Department of Agriculture, Washington, United States.
USDA (1999) Internet homepage of the Plant Genetic Data and Information Centre. U.S. Department of Agriculture, Agricultural Research System, National Agricultural Library (USDA/ARS), Washington, D.C,. United States. (http://www.nal.usda.gov/pgdic/Map_proj) (consulted January 1999).
Van der Have BV (1979) *100 Jaar Van der Have 1879-1979.* Koninklijk Kweekbedrijf en Zaadhandel, D.J. van der Have BV, Kapelle, the Netherlands.
Van der Pijl, K. (1984) *The Making of an Atlantic Ruling Class.* Verso, London, United Kingdom.
Van Harreveld, J. (1934) *Bescherming van den Kweekerseigendom.* J.H. de Bussy, Amsterdam, the Netherlands.
Van Soest, L.J.M. and Van Hintum, Th.J.L. (1995) *De Genenbank Onderweg: Verleden, Heden en Toekomst van het CGN.* CPRO-DLO, Wageningen, the Netherlands.
Van Wijk, J. (1990) Patents and the GATT. *Biotechnology and Development Monitor* 3, 24.
Van Wijk, J. (1997) The impact of intellectual property protection on seed supply. In: Tripp (ed.) (1997), pp. 185-197.
Van Wijk, J. (1998) Plant patenting provision reviewed in WTO. *Biotechnology and Development Monitor* 34, 6-9.
Van Wijk, J. and Jaffé, W. (eds.) (1996) Intellectual property rights and agriculture in developing countries. *Proceedings of a seminar on the impact of plant breeders' rights in developing countries, March 7-8, 1995, Bogotá, Colombia.* University of Amsterdam, Amsterdam, the Netherlands.

Van Wijk, J. and Junne, G. (1993) Intellectual property protection of advanced technology. Changes in the global technology system: implications and options for developing countries. INTECH Working Paper No. 10. United Nations University Institute for New Technologies, Maastricht, the Netherlands.

Vavilov, N.I. (1926) *Centri'i prodshozdenija kul'turn'ih rasternii* [Studies in the Origin of Cultivated Plants]. All-Union Institute of Applied Botany and New Crops, Leningrad, Soviet Union.

Vavilov, N.I. (1951) *The Origin, Variation, Immunity and Breeding of Cultivated Plants: Selected Writings of N.I. Vavilov.* Ronald Press Company, New York, United States.

Vavilov, N.I. (1992) The problem concerning the origin of cultivated plants. In: Vavilov N.I., *Origin and Geography of Cultivated Plants.* Cambridge University Press, Cambridge, United Kingdom. (Transcript of speech given at the All-Soviet Congress of Genetics, Plant Breeding, Agriculture and Livestock Breeding in Leningrad, January 10, 1929).

Vavilov, N.I. and Rodin, L.E. (1997) *Five Continents.* IPGRI, Rome, Italy.

Verfaillie, H. (1997) Speech of Monsanto president Hendrik Verfaillie to the Forum on Nature and Human Society at the National Academy of Sciences, October 30, 1997 Washington, D.C., United States. (http://www.monsanto.com/monsanto/media/docs/971030-verfaillie-h).

Verhulst, P.A. (1947) *Kweekersrecht.* Sijthoff's Uitgeversmaatschappij N.V., Leiden, the Netherlands.

Viatte, G. and Schmidhuber, J. (1998) Long-term policy issues and challenges for agro-food. In: OECD, *The Future of Food. Long-term Prospects for the Agro-Food Sector.* Organisation for Economic Co-operation and Development, Paris, France.

Vogel, J.H. (1994) *Genes for Sale: Privatization as a Conservation Policy.* Oxford University Press, Oxford, United Kingdom.

WBCSD (1995) International business leaders and Geneva officials inaugurate environmental business organization. Press releases of the World Business Council for Sustainable Development. WBCSD, Geneva, Switzerland.

WBCSD and IUCN (1997) *Business and Biodiversity: A Guide for the Private Sector.* World Business Council for Sustainable Development and the World Conservation Union, Gland, Switzerland.

Weibull, G.H. (1955) Plant breeders' rights. In: OEEC (1955), pp. 111-126.

White, A. (1964) The effect of breeders' rights on the seed trade. In: CSSA (1964), pp. 67-76.

Whyte, R.O. (1959) International approach to plant exploration and introduction. *Euphytica* 8, 159-208.

Wierema H. (1989) Seed industry development in developing countries. A survey amongst private seed companies in the North. Research report No.35, Development Research Institute (IVO), University of Tilburg, The Netherlands.

Wilkes, H.G. (1983) The current status of crop germplasm. *CRC Critical Reviews in Plant Sciences* 1(2), 133-181, pp. 337-341.

WIPO (1988) *Background Reading on Intellectual Property.* World Intellectual Property Organization, Geneva, Switzerland.

Witmeyer, D.M. (1994a) Organizing international cooperation: the international effort to conserve plant genetic resources. PhD thesis, University of Pittsburgh, Pittsburgh, United States.

Witmeyer, D.M. (1994b) The Convention on Biological Diversity changes rules of game for international plant genetic resources regime. *Diversity* 10(3) 28-31.

Witt, S.C. (1985) *Biotechnology and Genetic Diversity.* Briefbook Series. California

Agricultural Lands Project, California, United States.
Wolf, K.D. (1991) *Internationale Regime zur Verteilung Globaler Ressourcen.* Nomos Verlagsgesellschaft, Baden-Baden, Germany.
Wood, D. and Pistorius, R. (1993) The Global Environment Facility: sustainable value to developing countries? *Biotechnology and Development Monitor* 16, 10-11.
World Bank (1995) Chile: Estrategia para Elevar la Competitivdad Agrícola y Aliviar la Pobreza Rural. World Bank Study, World Bank, Washington, D.C., United States.
World Bank (1996) Review of Colombia's agriculture and rural development strategy. World Bank Country Study, World Bank, Washington, D.C., United States.
Yeatman, C.W., Kafton, D., Wilkes, G. (eds.) (1984) *Plant Genetic Resources: A Conservation Imperative.* AAAS Selected Symposium No. 87, Westview Press, Boulder, United States.
Young, O.R. (1989) The politics of international regime formation: managing natural resources and the environment. *International Organization* 43(3), 349-375.
Zeneca (1998) Zeneca Internet homepage. (http://www.zeneca.com) (consulted January 1999).
Zeven, A. (1996) Results of activities to maintain landraces and other material in some European countries *in situ* before 1945 and what we may learn from them. *Genetic Resources and Crop Evolution* 43, 337-341.

Appendix i

Persons Interviewed in Chile and Colombia

Chile

Accatino, P., Senior specialist, International Potato Center (CIP), Santiago, 28 March 1994.

Alamos, V., Vice-president, Agro Tuniche, Santiago, 16 December 1996.

Alessandri, G., Head Seed Division, Ciba Geigy, Chile, Santiago, 24 March 1994.

Anguita, A., Representative of Mapuche, Temuco, 19 December 1996.

Artigas, C., Legal advisor, Division de Medio Ambiente y Desarrollo, Comisión Económica Para America Latina y El Caribe (CEPAL), Santiago, 16 December 1996.

Baldrich, A., President, A.C. Baldrich-Chile, Santiago, 25 March 1994.

Benoit, I.L., Head, Section Wild Flora, Corporación National Forestal (CONAF), Santiago, 20 December 1996.

Concha, J.E., Director, Viveros Requinoa, Santiago, 16 December 1996.

Contreras, A., Facultad de Ciencias Agrarias, Universidad Austral de Chile, and President of the Asociación Chilana de la Papa (ACHIPA), Valdivia, 17 December 1996.

Cubillos, A., and P.L. Lobos, Programa de los Recursos Genéticos, Instituto Nacional de Investigación Agropecuaria (INIA), La Platina, 11 December 1996.

Cuchacovich, J.C., Head, Departamento Protección de Recursos Naturales Renovables, Servicio Agricola y Ganadero, Ministerio de Agricultura, Santiago, 19 December 1996.

Herrera, A., Instituto de Estudios Indígenas, Universidad de la Frontera, Temuco, 19 December 1996.

Heuchulaf, M., Corporación Nacional de la Desarollo des Indígenes (CONADI), 20 December 1996.

Izquierdo, J., Regional Plant Production Officer, FAO Latin American Regional Office,

Santiago, 10 December 1996.

Jordán, G., President, Instituto Nacional de Investigación Agropecuaria (INIA), Santiago, 16 December 1996.

Lara, A., Departamento de Silvicultura, Universidad Austral de Chile, Valdivia, 18 December 1996.

Lara, C., Consumers International, Santiago, 12 December 1996.

Legarraga, M., Director, Agricola LLahuen, Santiago, 24 March 1994.

Lerdon, J. and Salas, M., Instituto de Economía Agraria, Universidad Austral de Chile, Valdivia, 17 December 1996.

Lopez, H., Plant breeder, Instituto Nacional de Investigación Agropecuaria (INIA), La Platina, 13 December 1996.

Lorenz, M., Fundación Chile, Santiago, 12 December 1996.

Madrid, E., President, Asociación Nacional de Productores de Semillas (ANPROS), Santiago, 25 March 1994 and 10 December 1996.

Manzur, M.I., Coordinator Biodiversity Programme, Comité Nacional Pro Defensa de La Fauna y Flora, Santiago, 18 December 1996.

Melillan, M. and Antilef, L., Sociedad de Mapuche Newken, Temuco, 19 December 1996.

Messina, R., Head, Departamento de Semillas, Servicio Agrícola y Ganadero, Ministerio de Agricultura, Santiago, 12 December 1996.

Molina, A., President, Confederación Nacional de la Agricultura Familiar Campesina 'La Voz del Campo', Santiago, 12 December 1996.

Montesinos, C., Centro de Educación y Tecnología (CET), Santiago and Temuco, 23 March 1994 and 19 December 1996.

Muñoz, C., Coordinator, and Flores, L., Legal Advisor, Unidad de Patrinomio Ambiental, Commission Nacional del Medio Ambiente (CONAMA), Santiago, 18 December 1996.

Nain, J. and Santander, M., Consejo de Todas las Tierras, Temuco, 19 December 1996.

Orella, J., Sociedad Nacional de Agricultura (SNA), Santiago, 10 December 1996.

Parodi, P., Departamento de Ciencias Végétales, Universidad Católica de Chile, Santiago, 22 March 1994.

Portilla, B., Head Production and Commercialization, Instituto Nacional de Investigación

Agropecuaria (INIA), Santiago, 21 March 1994 and 10 December 1996.

Ramirez, I., wheat breeder, and Saavedra del Real, J., Agronomist, Instituto Nacional de Investigación Agropecuaria (INIA), La Platina, 11 December 1996.

Reyes, B., Marcelo, W. and Hernting, H., Instituto de Ecología Política, Santiago, 20 December 1996.

Saenger, R., Seed Division Manager, Cargill Chile, Rancagua, 23 March 1994.

Salvo, B., Agronomist, Comité de Obtentores de Nuevas Variedades Végétales, Asociación Nacional de Productores de Semillas (ANPROS), Santiago, 10 December 1996.

Seemann, P., Facultad de Ciencias Agrarias, Universidad Austral de Chile, Valdivia, 18 December 1996.

Simonetti, J., Faculty of Ecology, Universidad Católica de Chile, Santiago, 10 December 1996.

Schmidt, A., General Director, and M. Bacigalupo, Commercial Director, Experimental Station, Sociedad Nacional de Agricultura (SNA), Santiago, 23 March 1994

Villarroel, P., Editor, Centro de Investigación y Planificación del Medio Ambiente (CIPMA), Valdivia, 17 December 1996.

Von Baer, E., Director, Von Baer Semillas, Santiago, 25 March 1994 and 24 May 1996.

Colombia

Aguilar, J.A. and Castaño, G., Instituto Major Campesino (IMCA), Buga, 13 February 1994.

Ahumada, J., Director, PROHOSA Ltda, Bogotá, 10 February 1994.

Alvarado, L.R., Director, Ecotecnica Ltda, Bogotá, 7 November 1996.

Anzolla M., Subdirección de Relaciones de Integración y Organismos Multilaterales, Ministerio de Comercio Exterior, Bogotá, 14 November 1996.

Aramendis, R., Programa Nacional de Biotecnología, Colciencias, Bogotá, 18 November 1996.

Arango, M., Royalty Administration International, Bogotá, 12 November 1996.

Arevalo, H. and Luján, L., Federación Colombiana de Productores de Papa (FEDEPAPA), Bogotá, 18 November 1996.

Arthehortúa, L., Departamento de Biología, Universidad de Antioquia, Medellín, 15 November 1996.

Barbosa, J., Director, Pioneer Hi-bred Colombia, Cali, 15 February 1994.

Camacho, M., General Director, Floramerica, Bogotá, 17 February 1994.

Casas, A., Consultant, Ministerio del Medio Ambiente, Bogotá, 22 November 1996.

Cock, J., General Director, Centro de Investigación en Caña Azúcar (CENICANA), Cali, 14 February 1994.

Cogollo, A., Fundación Jardín Botánico "Joaquin Antonio Uribe", Medellín, 15 November 1996.

Copete, E., Director, Ciba Geigy Colombia, Cali, 15 February 1994.

Debouck, D., Head of Genebank, International Center for Tropical Agriculture (CIAT), Cali, 26 November 1996.

Flóres, M., Instituto Latinamericano de Servicios Legales Alternativos (ILSA), Bogotá, 22 November 1996.

Fonseca, A., Member of Congress, Bogotá, 21 November 1996.

Gómez, R.A., President, Asociación Nacional de Usarios Campesinos (ANUC), Bogotá, 12 November 1996,

Green, A., Organización Nacional Indígena de Colombia (ONIC), Bogotá, 28 November 1996.

Herron, H., Research director, Federación Nacional de Cafeteros (FEDECAFE), Bogotá, 28 November 1996.

Isaacs, M., Consultant, Corporación Colombia Internacional (CCI), Bogotá, 2 December 1996.

Laignelet, A., Head R&D, Meristemos, Bogotá, 30 May 1994.

La Rota, R., President, Propagar Plantas, Bogotá, 1 June 1994.

Lobo, M., National coordinator plant genetic resources, Corporación Colombiana de Investigación Agropecuario (CORPOICA), Bogotá, 13 November 1996.

Martínez, P., Legal Advisor, Organización Indígena de Antioquia (OIA), Medellín, 15 November 1996.

Michelsen E., Vice-president, and Piñeda, C., Executive Director, Fundación Para la Defensa del Interes Publico (FUNDEPUBLICO), Bogotá, 14 November 1996.

Moscardi, E. Director, Inter-American Institute for Cooperation on Agriculture, Bogotá, 8 November 1996.

Muelas, L., Senator Colombian Congress, Bogotá, 28 November 1996.

Muñoz, D., Head plant breeding, Federación de Productores de Arroz (FEDEARROZ), Bogotá, 11 February 1994.

Nemoga G.R., Director, Unidad de Investigaciones JurídicoSociales "Gerardo Molina" (Unijus), Facultad de Derecho, Universidad Nacional, Bogotá, 8 November 1996.

Patiño, M.I., Director, Asociación Colombiana de Exportadores de Flores (ASOCOLFLORES), Bogotá, 9 February 1994.

Pombo, D., Instituto de Gestión Ambiental, Bogotá, 29 November 1996.

Quintama, A., President, and Pordnía, J., Federación Nacional de Cerealistas (FENALCE), Bogotá, 18 November 1996.

Roca, W., Head Biotechnology Unit, International Center for Tropical Agriculture (CIAT), Cali, 26 November 1996.

Rueda, H., Ministerio de Agricultura y Desarrollo Rural, Bogotá, 20 November 1996.

Sanabria, E.A. and A. Dabalos, Seeds Division, Federación de Productores de Arroz (FEDEARROZ), Bogotá, 19 November 1996.

Silva, C., Coordinator, Unidad de Seguimiento y Evaluación, Instituto Colombiano Agropecuario (ICA), Bogotá, 13 November 1996.

Silva, G., General Director, Semillas Valle, Cali, 16 February 1994.

Spijkers, P., private consultant, Bogotá, 7 November 1996.

Suarez, J., Head, and A.L. Diaz and A. Santanilla, Seeds Division, Instituto Colombiano Agropecuario (ICA), Bogotá, 13 November 1996.

Torres, R., Director Research, Corporación Colombiana de Investigación Agropecuario (CORPOICA), Bogotá, 9 February 1994 and 21 November 1996.

Urres, R., Programa Regional Agrícola, Corporación Colombiana de Investigación Agropecuario (CORPOICA), Bogotá, 13 November 1996.

Veléz, G., Agronomist, Programa Semillas, Bogotá, 13 November 1996.

Villota, E., Director, Asociación de Productores de Semillas (ACOSEMILLAS), Bogotá, 19 November 1996.

Williams, D., Interim Regional Director, International Plant Genetic Resources Institute (IPGRI), Regional Office for the Americas, Cali, 25 November 1996.

Appendix ii

Note on the Interviews

During the fieldwork in Chile and Colombia, 102 persons representing around 75 organizations or departments were interviewed (Figure I). Of the total number of 91 interviews, 23 were carried out in 1994 and 68 in 1996. The interviewees in Chile and Colombia were basically selected by means of the "snowball" method, which means that a small group of persons who were involved in crop development were identified, interviewed, and used as informants to identify others who qualified for inclusion in our sample. This method of sampling was considered to be appropriate as it met our aim to provide an overview of the various opinions on plant-related conservation and intellectual property protection. As many of the interviewees represented a specific social category, we were able to relate their opinions to the specific position of this category in the agro-industrial production process.

We tried very actively to include in the sample all individuals and organizations that had previously expressed an opinion on biodiversity conservation and plant-related intellectual property protection. We interviewed representatives of most governmental departments and institutions that were responsible for policy in this field. As far as the views of farmers and indigenous peoples are concerned, we relied on the position of the main organizations that represent these groups. We also interviewed the main associations of the crop development industry, as well as a significant number of individual companies. Most, if not all, public and private plant breeding organizations in the two countries were included in the sample. Added to the sample were specialists from universities, NGOs, and international agricultural research organizations who either played a direct role in the design of biodiversity conservation or intellectual property policy, or who appeared to have a specific interest in it.

Two additional comments on the categorization of interviewees should be made:
a) Some organizations are involved in more than one activity. For example, several farmers' associations also have their own plant breeding programmes. In such cases the interviewees are categorized according to the department they were working for,

i.e. representing either a breeding organization, or a farmers' interest association.
b) A private firm is considered to be domestic when 51% or more of the capital is owned by national shareholders. Foreign companies are those companies of which 51% of the capital is owned by foreign shareholders. Because data about the companies' shares are not publicly available, it was left to the respondent to indicate whether or not a company is considered to be national or foreign according to our criterion.

Figure I Number of interviewees by country and type of organization

		Breeder				Association								
	Government[1]	Private D[2]	Private F[3]	Public[4]	Propagator[5]	Cultivator[6]	Industry[7]	Farmers[8]	NGO[9]	Indigenous[10]	International[11]	University[12]	Other[13]	Total
Chile	9	3	2	4	2	1	3	2	7	6	3	7	1	50
Colombia	10	4	2	3	6	1	1	5	7	2	4	2	5	52
Total	19	7	4	7	8	2	4	7	14	8	7	9	6	102

[1] Government, including both Members of Parliament and officials from ministries in the area of agriculture, industry, forestry, and environmental protection
[2] Private domestic seed company with plant breeding activities
[3] Private foreign seed company with plant breeding activities
[4] Public institute with plant breeding activities
[5] Private domestic seed company with propagation activities
[6] Private domestic cultivating companies
[7] Associations of breeders, propagators, or cultivators
[8] Associations of farmers
[9] Non-governmental organizations in the area of nature protection or small-scale farming
[10] Organizations of indigenous peoples
[11] International organizations, including: CIAT, FAO, IPGRI, IICA
[12] University departments with research on agriculture or biodiversity conservation
[13] Other organizations with an interest in crop development

Index